Name ______________________________ Class ________

Skills Worksheet

Directed Reading

Section: What Is an Ecosystem?

In the space provided, write the letter of the description that best matches the term or phrase.

______ **1.** ecology

______ **2.** habitat

______ **3.** community

______ **4.** ecosystem

______ **5.** abiotic factors

______ **6.** biotic factors

a. the place where a population of a species lives

b. a community and all of the physical aspects of its habitat

c. the organisms living in a habitat

d. the physical aspects of a habitat

e. the different species that live in a habitat

f. the study of the interactions of living organisms with one another and with their physical environment

In the space provided, write the letter of the term or phrase that best completes each statement or best answers each question.

______ **7.** An ecosystem is defined by the
- **a.** organisms living in a particular area.
- **b.** climate of the region.
- **c.** community and physical aspects of the area.
- **d.** pioneer species that occupy the area.

______ **8.** A measure of how many different species live in an ecosystem is its
- **a.** ecology.
- **b.** biodiversity.
- **c.** biology.
- **d.** abiotic factor.

______ **9.** A typical organism that might occupy a forest in the southeastern United States is a
- **a.** cactus.
- **b.** wolf.
- **c.** deer.
- **d.** polar bear.

______**10.** Which area could you expect to find the most biodiversity?
a. a square yard of forest floor
b. a square yard of desert sand
c. the surface of a mountain stream
d. a rock recently exposed by a retreating glacier

______**11.** An ecosystem includes which of the following?
a. microscopic organisms
b. large mammals
c. weather conditions
d. All of the above

______**12.** Which of the following can be considered an entire ecosystem?
a. a small patch of farmland
b. the underside of a large rock
c. a stream flowing through a pasture
d. All of the above

______**13.** An example of a biotic factor in an ecosystem is
a. an active volcano.
b. the number of competing species.
c. a river that floods frequently.
d. the intensity of sunlight in summer.

Read each question, and write your answer in the space provided.

14. What is meant by the term *pioneer species*?

____

____

15. Explain the differences among succession, primary succession, and secondary succession.

____

____

____

____

16. Why is Glacier Bay, Alaska, an example of how ecosystems change over time?

____

____

____

____

Name ______________________ Class ______________ Date ____________

Skills Worksheet

Directed Reading

Section: Energy Flow in Ecosystems

In the space provided, explain how the terms in each pair differ in meaning.

1. producers, consumers

____

____

____

____

2. trophic level, food chain

____

____

____

____

3. herbivores, carnivores

____

____

____

____

4. detritivores, decomposers

____

____

____

____

In the space provided, write the letter of the description that best matches the term or phrase.

______ 5. omnivore

6. herbivore

______ 7. producer

______ 8. detritivore

______ 9. decomposer

______ 10. consumer

______ 11. carnivore

______ 12. food web

______ 13. food chain

a. interconnected group of food chains

b. cause decay

c. a path of energy through the trophic levels of an ecosystem

d. eat only plants

e. eat only animals

f. organisms that first capture energy

g. eat both plants and animals

h. consume plants or other organisms to obtain energy

i. obtain energy from organic wastes and dead bodies

Complete each statement by writing the correct term or phrase in the space provided.

14. At each trophic level, the energy stored is about ______________ percent of that stored by the organisms in the level below.

15. A(n) ______________ ______________ is a diagram in which each trophic level is represented by a block.

16. The rate at which organic material is produced by photosynthetic organisms in an ecosystem is called ______________

______________ .

Name ______________________ Class ______________ Date ____________

Skills Worksheet

Directed Reading

Section: Cycling of Materials in Ecosystems

Read each question, and write your answer in the space provided.

1. What are biogeochemical cycles?

____

____

____

2. What are living and nonliving reservoirs?

____

____

____

3. What are the most important substances that pass through biogeochemical cycles?

____

Complete each statement by underlining the correct term or phrase in the brackets.

4. In a tropical rain forest, most of the water in the atmosphere comes from [evaporation / transpiration].

5. Water that falls to the Earth as rain or snow and seeps into the soil becomes [surface / ground] water.

6. In the living portion of the water cycle, water is taken up by [condensation / the roots of plants].

7. The process by which water evaporates from the leaves of plants is called [respiration / transpiration].

Read each question, and write your answer in the space provided.

8. How does carbon become part of organic molecules?

____

____

____

9. List three ways carbon atoms return to the nonliving reservoir.

____

____

____

Complete each statement by writing the correct term or phrase in the space provided.

10. Organisms need nitrogen and phosphorus to build ______________

and ______________ ______________.

11. Phosphorus is usually present as ______________

______________ in soil and rock.

12. The process of combining nitrogen gas with hydrogen to form ammonia is

called ______________ ______________.

13. Nitrogen-fixing bacteria use ______________ to split molecules of nitrogen gas and combine the nitrogen atoms with hydrogen.

Name ______________________________ Class ________________ Date ____________

Skills Worksheet

Active Reading

Section: What Is an Ecosystem?

Read the passage below. Then answer the questions that follow.

Ecology is the study of the interactions of living organisms with one another and with their physical environment. The place where a particular population of a species lives is its **habitat.** The many different species that live together in a habitat are called a **community.** An **ecosystem**, or ecological system, consists of a community and all the physical aspects of its habitat, such as the soil, water, and weather. The physical aspects of a habitat are called **abiotic factors,** and the living organisms in a habitat are called **biotic factors.** The number of species living within an ecosystem is a measure of its **biodiversity.**

SKILL: READING EFFECTIVELY

In the space provided, write the term or phrase from the passage above that best matches the description. Some terms or phrases may be used more than once.

______________________ **1.** all living organisms in a habitat

______________________ **2.** number of species living within an ecosystem

______________________ **3.** study of a habitat's abiotic and biotic factors

______________________ **4.** deer, squirrels, and rabbits living together in a forest form this

______________________ **5.** an ecological system

______________________ **6.** soil, water, and weather are examples of these

______________________ **7.** place where a population lives

______________________ **8.** all species of freshwater fish that live together in a lake form this

______________________ **9.** consists of a community and abiotic factors

An analogy is a comparison. In the space provided, write the letter of the term that best completes the analogy.

______ **10.** Biotic is to bird as abiotic is to

a. grass.
b. tree.
c. nest.
d. worm.

Name ______________________ Class ______________ Date ____________

Skills Worksheet

Active Reading

Section: Energy Flow in Ecosystems

Read the passage below. Then answer the questions that follow.

Ecologists study how energy moves through an ecosystem by assigning organisms in that ecosystem to a specific level, called a **trophic level,** in a graphic organizer based on the organism's source of energy.

Energy moves from one trophic level to another. The path of energy through the trophic levels of an ecosystem is called a **food chain.** The lowest trophic level of any ecosystem is occupied by the producers, such as plants, algae, and bacteria. Producers use the energy of the sun to build energy-rich carbohydrates. Many producers also absorb nitrogen gas and other key substances from the environment and incorporate them into their biological molecules.

At the second trophic level are **herbivores,** animals that eat plants and other primary producers. They are the primary consumers. Cows and horses are herbivores, as are caterpillars and some ducks. A herbivore must be able to break down a plant's molecules into usable compounds. However, the ability to digest cellulose is a chemical feat that only a few organisms have evolved. As you will recall, cellulose is a complex carbohydrate found in plants. Most herbivores rely on microorganisms, such as bacteria and protists, in their gut to help digest cellulose. Humans cannot digest cellulose because we lack these particular microorganisms.

At the third trophic level are secondary consumers called **carnivores,** animals that eat herbivores. Tigers, wolves, and snakes are carnivores. Some animals, such as bears, are both herbivores and carnivores; they are called **omnivores.** They use the simple sugars and starches stored in plants as food, but they cannot digest cellulose.

SKILL: READING EFFECTIVELY

Read each question, and write your answer in the space provided.

1. What relationship exists between trophic levels and a food chain?

____

____

2. What group of organisms occupies the first trophic level of an ecosystem?

____

3. What group of organisms occupies the second trophic level of an ecosystem?

____

____

4. What group of organisms occupies the third trophic level of an ecosystem?

____

____

5. How are omnivores similar to carnivores? How do they differ?

____

____

SKILL: SEQUENCING INFORMATION

In the space provided, write *1* if the statement describes the first trophic level, write *2* if the statement describes the second trophic level, or write *3* if the statement describes the third trophic level.

______ **6.** Primary consumers are found here.

______ **7.** Organisms here use the energy of the sun to build energy-rich carbohydrates.

______ **8.** Tigers, wolves, and snakes are found here.

______ **9.** Organisms here are capable of breaking down cellulose.

______ **10.** Secondary consumers are found here.

______ **11.** Plants, algae, and bacteria are found here.

______ **12.** Humans are found here.

______ **13.** Organisms here break down a plant's molecules into usable compounds.

In the space provided, write the letter of the term or phrase that best completes the statement.

______ **14.** All of the following are examples of primary consumers EXCEPT

a. maple trees.
b. caterpillars.
c. cows.
d. horses.

Name ______________________ Class ______________ Date ____________

Skills Worksheet

Active Reading

Section: Cycling of Materials in Ecosystems

Read the passage below. Then answer the questions that follow.

In the nonliving portion of the water cycle, water vapor in the atmosphere condenses and falls to the Earth's surface as rain or snow. Some of this water seeps into the soil and becomes part of the **ground water,** which is water retained beneath the surface of the Earth. Most of the remaining water that falls to the Earth does not stay at the surface. Instead, heated by the sun, it reenters the atmosphere by evaporation.

In the living portion of the water cycle, much water is taken up by the roots of plants. After passing through a plant, the water moves into the atmosphere by evaporating from the leaves, a process called **transpiration.** Transpiration is also a sun-driven process. The sun heats the Earth's atmosphere, creating wind currents that draw moisture from the tiny openings in the leaves of plants.

SKILL: READING EFFECTIVELY

Read each question, and write your answer in the space provided.

1. What occurs in the nonliving part of the water cycle?

____

____

2. What happens to this precipitation?

____

____

3. What occurs in the living part of the water cycle?

____

____

4. What is transpiration?

____

____

5. Why is transpiration classified as a "sun-driven process"?

____

____

____

SKILL: INTERPRETING GRAPHICS

The figure below shows the water cycle. Insert the following terms in the correct spaces: *evaporation, ground water, precipitation, transpiration,* and *water vapor*.

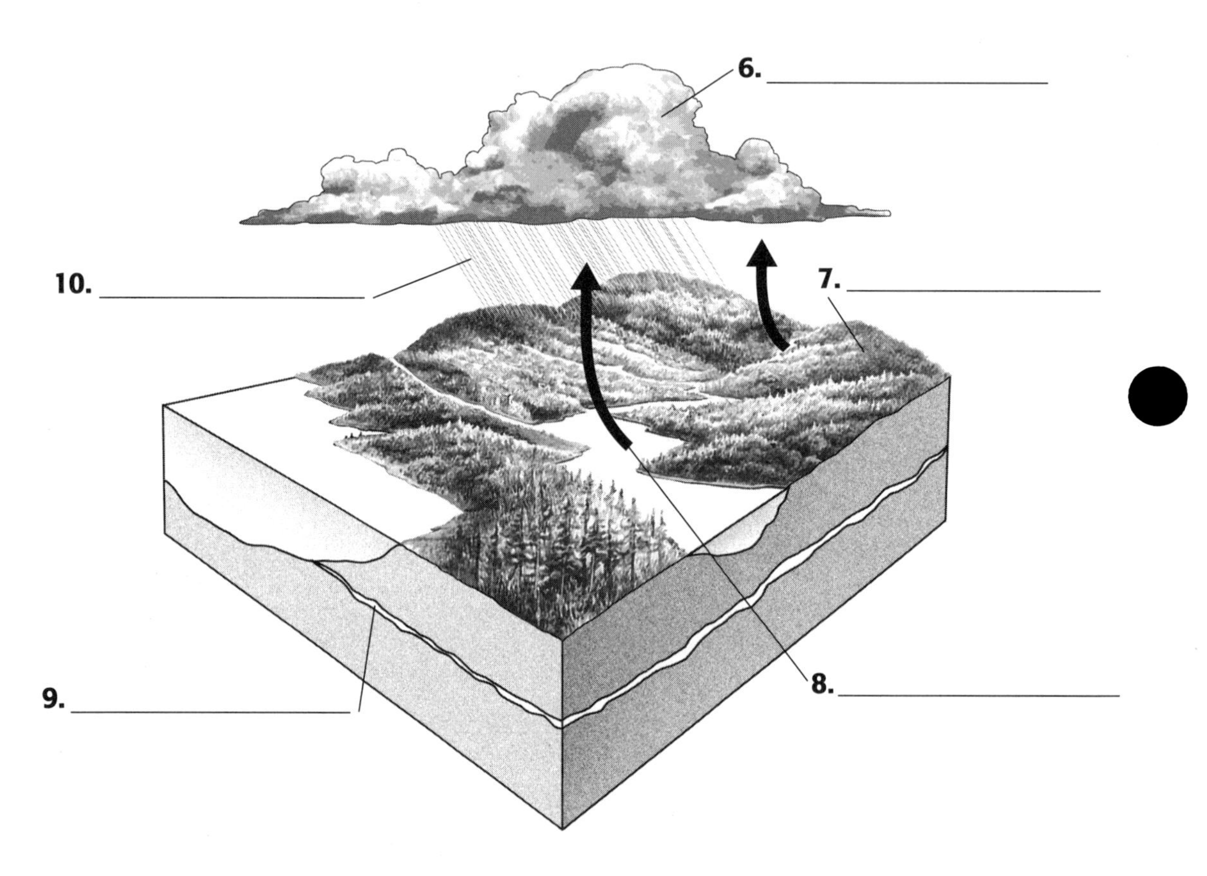

In the space provided, write the letter of the term that best completes the statement.

______ **11.** Water retained beneath Earth's surface is called

a. precipitation.
b. transpiration.
c. condensation.
d. ground water.

Name ______________________ Class ______________ Date ____________

Skills Worksheet

Vocabulary Review

Complete each statement by writing the correct term or phrase from the list below in the space provided.

abiotic factors	ecology	primary succession
biodiversity	ecosystem	secondary succession
biotic factors	habitat	succession
community	pioneer species	

1. The number of species living within an ecosystem is a measure of its

______________________ .

2. A somewhat regular progression of species replacement is called

______________________ .

3. A(n) ______________________ consists of a community and all the physical aspects of its habitat, such as the soil, water, and weather.

4. The living organisms in a habitat are called ______________________

______________________ .

5. The first organisms to live in a new habitat are small, fast-growing plants

called ______________________ ______________________ .

6. Succession that occurs where plants have not grown before is

called______________________ ______________________ .

7. The many different species that live together in a habitat are called

a(n) ______________________ .

8. The study of the interactions of living organisms with one another and

with their environment is called ______________________ .

9. Succession that occurs where previous growth has occurred is called

______________________ ______________________ .

10. The physical aspects of a habitat are called ______________________

______________________ .

11. The place where a particular population of a species lives is called its

______________________ .

In the space provided, write the letter of the description that best matches the term or phrase.

______ **12.** primary productivity

______ **13.** producers

______ **14.** consumers

______ **15.** trophic level

______ **16.** food chain

______ **17.** herbivore

______ **18.** carnivore

______ **19.** omnivore

______ **20.** detritivore

______ **21.** decomposers

______ **22.** food web

______ **23.** energy pyramid

______ **24.** biomass

______ **25.** biogeochemical cycle

______ **26.** ground water

______ **27.** transpiration

______ **28.** nitrogen fixation

a. an interconnected group of food chains

b. a pathway formed when a substance enters a living organism, stays for a time in the organism, then returns to the nonliving environment

c. the dry weight of tissue and other organic matter found in a specific ecosystem

d. organisms in an ecosystem that first capture energy

e. water retained beneath the surface of Earth

f. the rate at which organic material is produced by photosynthetic organisms

g. a diagram in which each trophic level is represented by a block with a width proportional to the amount of energy stored in the organisms at that trophic level

h. the process of combining nitrogen with hydrogen to form ammonia

i. organisms that obtain energy by consuming plants or other organisms

j. the evaporation of water from the leaves of plants

k. a level in a diagram based on the organism's source of energy

l. an organism that obtains energy from organic wastes and dead bodies

m. the path of energy through the trophic levels of an ecosystem

n. bacteria and fungi that cause decay

o. an animal that is both a herbivore and a carnivore

p. an animal that eats other animals

q. an animal that eats plants or other primary producers

Name ______________________ Class ______________ Date ____________

Skills Worksheet

Science Skills

Interpreting Graphics

Use the figure below, which shows the food web of an aquatic ecosystem, to complete items 1–7.

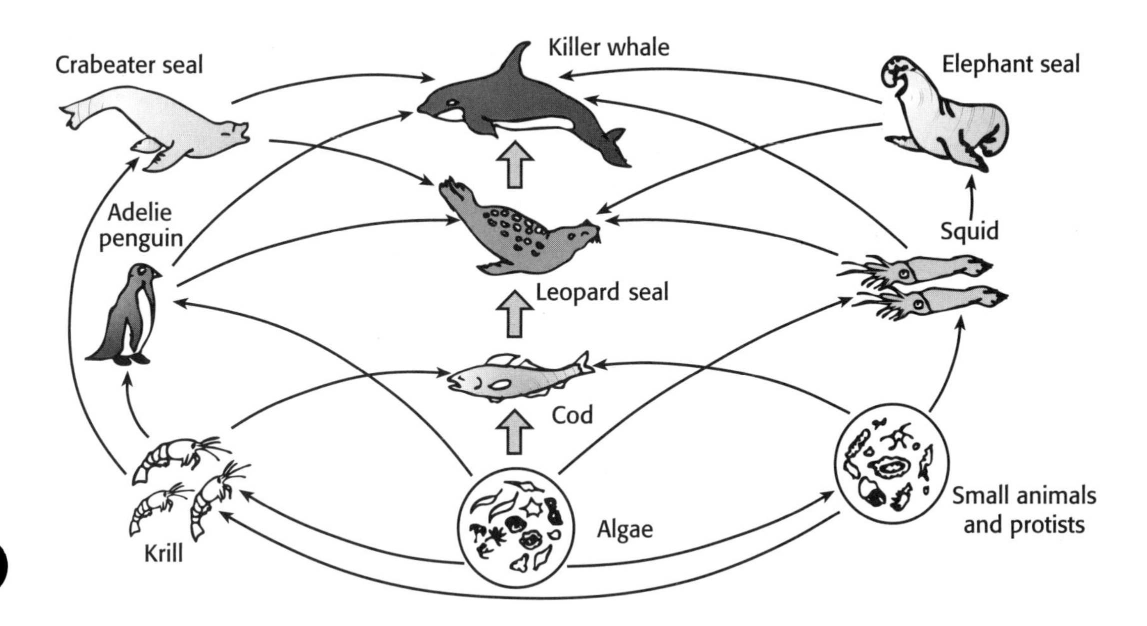

In the food web above, there are eight food chains that include krill. In the space provided, identify all of the organisms in the order in which they occur in four of these eight food chains.

1. Chain 1 ____

____

2. Chain 2 ____

____

3. Chain 3 ____

____

4. Chain 4 ____

____

Read each question about the food web on the previous page, and write your answer in the space provided.

5. What organisms do cod eat?

____

____

6. List all the organisms that eat squid.

____

____

7. How many producers are in the food web? Name them.

____

____

Use the figures below, which show trophic levels in an ecosystem, to complete items 8–11.

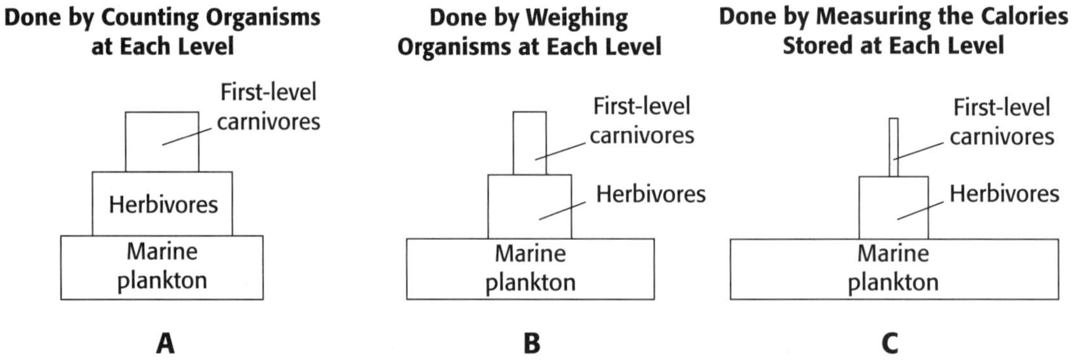

Done by Counting Organisms at Each Level

First-level carnivores
Herbivores
Marine plankton

A

Done by Weighing Organisms at Each Level

First-level carnivores
Herbivores
Marine plankton

B

Done by Measuring the Calories Stored at Each Level

First-level carnivores
Herbivores
Marine plankton

C

Study the three pyramids above. In the space provided, identify which pyramid is the most accurate indicator of each item below by writing the correct letter (*A–C*) in the space provided.

______ **8.** number of individual organisms

______ **9.** measurement of productivity

______ **10.** measurement of biomass

11. Which pyramid is the most accurate indicator of the amount of energy available at each trophic level? Explain.

____

____

____

Name ______________________ Class ______________ Date ______________

Skills Worksheet

Concept Mapping

Using the terms and phrases provided below, complete the concept map showing the characteristics of ecosystems.

abiotic factors	energy	nitrogen	soil
biotic factors	food chains	primary succession	succession
carbon	food webs	secondary succession	water

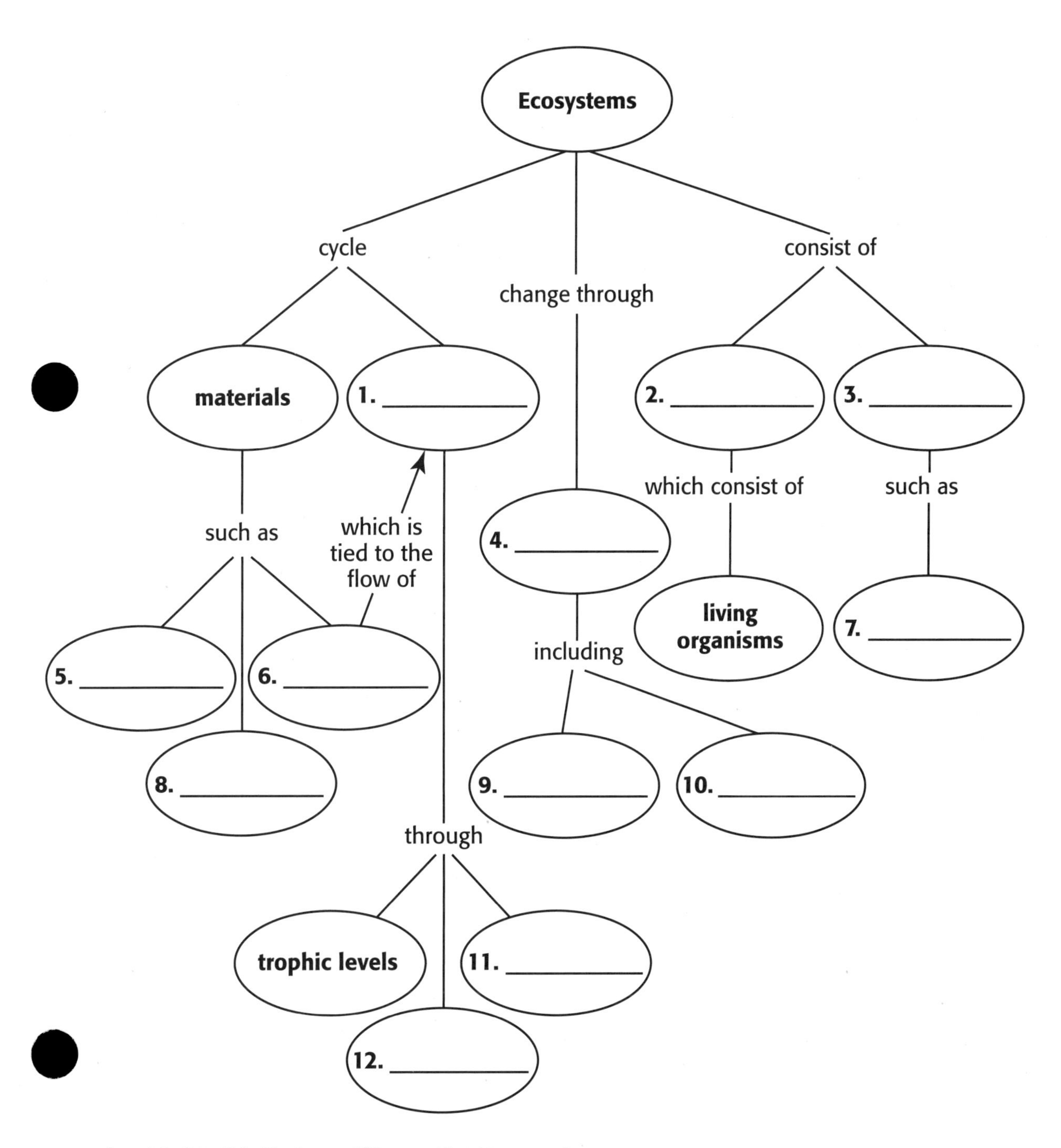

Name ______________________ Class ______________ Date ____________

Skills Worksheet

Critical Thinking

Work-Alikes

In the space provided, write the letter of the term or phrase that best describes how each numbered item functions.

______ **1.** all of Earth's organisms

______ **2.** biodiversity

______ **3.** energy lost at trophic levels

______ **4.** ammonification

a. heat produced by an internal combustion engine

b. ammonia factory

c. salad bar

d. jigsaw puzzle pieces

Cause and Effect

In the space provided, write the letter of the term or phrase that best matches each cause or effect given below.

Cause	Effect
5. removal of every organism from an ecosystem	______________
6. ______________	nutrients released back into the environment
7. ______________	the number of trophic levels in an ecosystem is limited
8. photosynthesis	______________

a. ecosystems lose energy at each trophic level

b. minerals, organic compounds, water, and sunlight remain

c. organic molecules

d. decomposers

Trade-offs

In the space provided, write the letter of the bad news item that best matches each numbered good news item below.

Good News

______ **9.** Succession brings in new species of plants.

______ **10.** 10 kg of grain eaten builds 1 kg of human tissue.

______ **11.** Water, carbon, nitrogen, and phosphorus pass from nonliving things to living organisms.

______ **12.** The atmosphere is 79 percent nitrogen.

Bad News

a. 100 kg of consumed beef required to build 1 kg of human tissue.

b. Most organisms cannot use it in the form of a gas.

c. Many of the old ones die due to the replacement.

d. There are less of these compounds in living reservoirs than in nonliving reservoirs.

Linkages

In the spaces provided, write the letters of the two terms or phrases that are linked together by the term or phrase in the middle. The choices can be placed in any order.

13. ______ raccoons, foxes, rabbits, chipmunks ______

14. ______ shrubs, grasses, small trees ______

15. ______ alder thicket ______

16. ______ herbivores ______

17. ______ cellulose-digesting bacteria in stomachs ______

a. worms, spiders, snakes, toads

b. producers

c. pine trees

d. mosses, fireweed, cottonwood, willows, and *Dryas*

e. Sitka spruce and hemlock trees

f. bears and white-tailed deer

g. cellulose is digested

h. carnivores

i. algae, bacteria, and fungi

j. herbivores eat plant material

Analogies

An analogy is a relationship between two pairs of terms or phrases written as a : b :: c : d. The symbol : is read as "is to," and the symbol :: is read as "as." In the space provided, write the letter of the pair of terms or phrases that best completes the analogy shown.

______**18.** food web : food chains ::
- **a.** ecosystem : a community
- **b.** bacterium : plants
- **c.** plant : communities
- **d.** protist : animals

______**19.** hawk : biotic factor ::
- **a.** spider : abiotic factor
- **b.** water : abiotic factor
- **c.** weather : biotic factor
- **d.** soil : biotic factor

______**20.** primary succession : areas of no plant growth ::
- **a.** primary succession : hemlock forest
- **b.** lichens : alders
- **c.** alder forests : bare rock
- **d.** secondary succession : forest clearing

______**21.** mosses and willows : alders ::
- **a.** succession : pioneer species
- **b.** mosses and willows : bare rock
- **c.** bare rock : pioneer species
- **d.** alders : bare rock

______**22.** plant : sunlight into chemical energy ::
- **a.** animal : sunlight into chemical energy
- **b.** plant : chemical energy into sunlight
- **c.** chemical energy : sunlight into nitrogen
- **d.** bacterium : ammonia into nitrates

Name ______________________ Class ______________ Date ______________

Skills Worksheet

Test Prep Pretest

In the space provided, write the letter of the term or phrase that best completes each statement or best answers each question.

______ **1.** Biodiversity is the number of species
- **a.** of animals living within an ecosystem.
- **b.** of plants and fungi living within an ecosystem.
- **c.** of bacteria and protists living within an ecosystem.
- **d.** living within an ecosystem.

______ **2.** The plants that first grow on an island formed by a volcano are part of a progression called
- **a.** primary succession.
- **b.** primary productivity.
- **c.** secondary succession.
- **d.** the climax community.

______ **3.** In the living portion of the water cycle, water
- **a.** is retained beneath the surface of Earth as ground water.
- **b.** evaporates from the soil.
- **c.** evaporates from dead organisms.
- **d.** is taken up by the roots of plants.

Questions 4–7 refer to the figure at right.

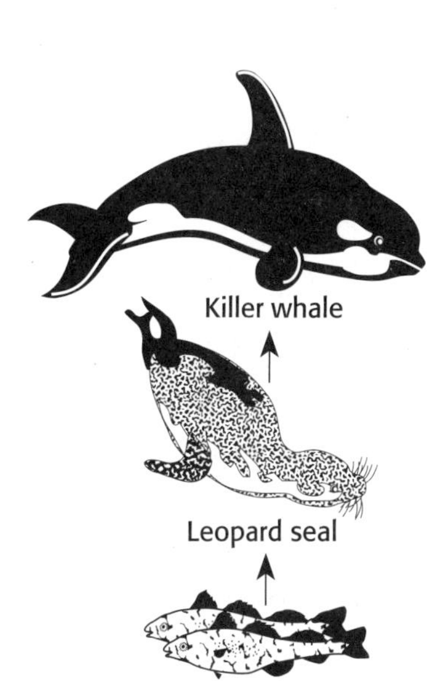

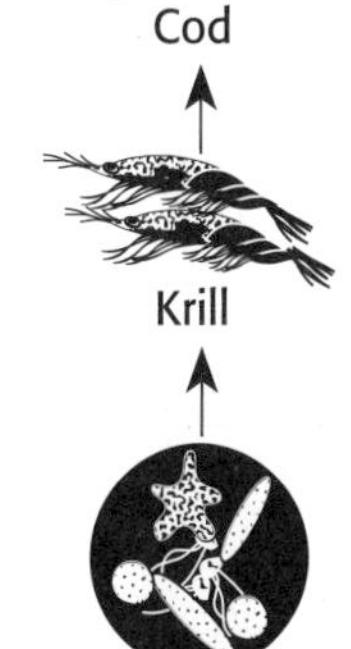

______ **4.** The algae are
- **a.** decomposers.
- **b.** consumers.
- **c.** producers.
- **d.** herbivores.

______ **5.** The krill are
- **a.** decomposers.
- **b.** consumers.
- **c.** producers.
- **d.** detritivores.

______ **6.** This figure is called a
- **a.** food chain.
- **b.** food web.
- **c.** pyramid of energy.
- **d.** trophic level.

______ **7.** The most likely reason that this figure shows only five levels is that
- **a.** pollution probably destroyed all of the higher levels.
- **b.** no other organisms are powerful enough to kill and eat the killer whale.
- **c.** too much energy is lost at each level to permit more levels.
- **d.** there is not enough energy initially present at the first level.

______ **8.** The process of succession varies depending on
- **a.** the plant species involved.
- **b.** initial environmental conditions and chance.
- **c.** pioneer species.
- **d.** competition between species.

______ **9.** The conversion of nitrate to nitrogen gas is called
- **a.** assimilation.
- **b.** ammonification.
- **c.** nitrification.
- **d.** denitrification.

In the space provided, write the letter of the description that best matches the term or phrase.

______ **10.** habitat

______ **11.** community

______ **12.** ecosystem

______ **13.** herbivores

______ **14.** carnivores

- **a.** animals at the second trophic level that eat plants
- **b.** the place where a particular population of a species lives
- **c.** the many species that live together in a habitat
- **d.** animals at the third trophic level that eat other animals
- **e.** a community and all the physical aspects of its habitat

Complete each statement by writing the correct term or phrase in the space provided.

15. The physical aspects, or ______________ ______________ , of an ecosystem's habitat include soil, water, and weather.

16. In a(n) ______________ ______________ , the amount of energy stored at each level determines the width of each block.

17. The amount of energy in a trophic level is more accurately determined by measuring the ______________ (dry weight of tissue) than the ______________ of organisms.

18. The process of combining nitrogen gas with hydrogen to form ammonia is called ______________ ______________ .

19. The production of ammonia by bacteria during the decay of nitrogen-containing urea is called ____________________ .

Read each question, and write your answer in the space provided.

20. What components are included in an ecosystem but not in a community?

____

____

____

21. Why are energy pyramids never inverted?

____

____

____

22. Trace the cycling of water between the atmosphere and Earth.

____

____

____

____

23. List the four stages of the nitrogen cycle.

____

____

____

Questions 24 and 25 refer to the figure below, which shows the carbon cycle.

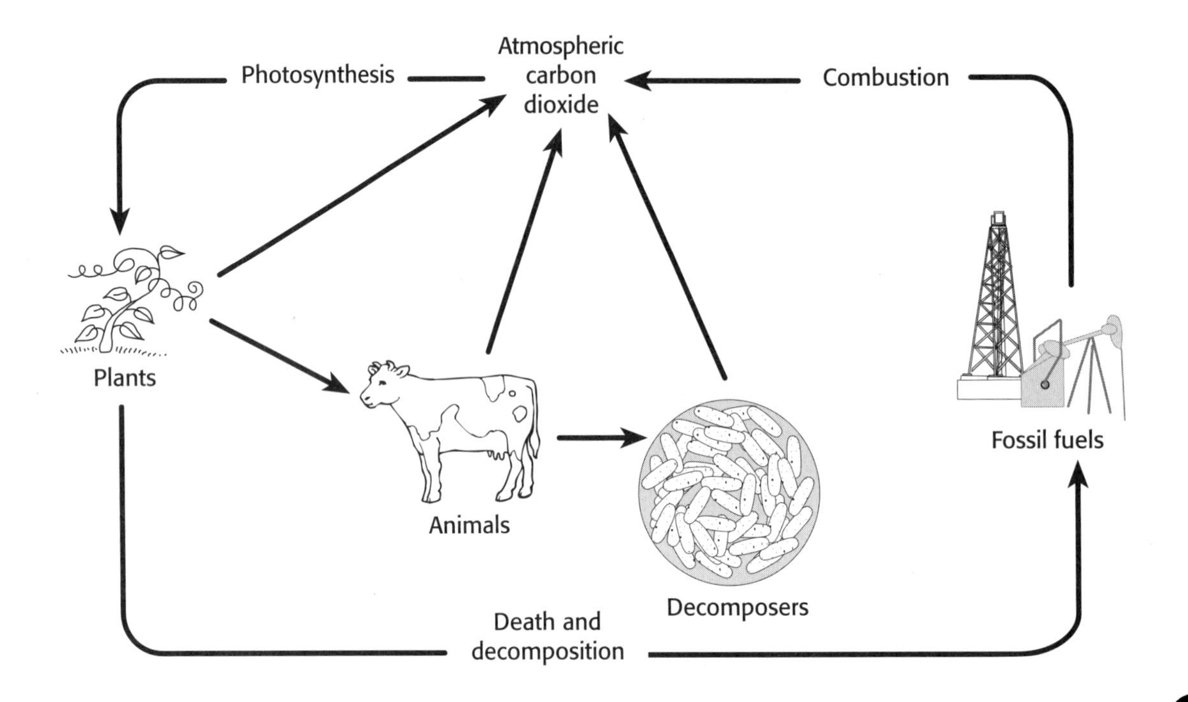

24. How do the living organisms in the figure return carbon atoms to the pool of carbon dioxide in the atmosphere and water?

____

____

____

25. What is the source of the carbon in fossil fuels?

____

____

____

Name ______________________________ Class ______________ Date ____________

Assessment

Quiz

Section: What Is an Ecosystem?

In the space provided, write the letter of the term or phrase that best completes each statement or best answers each question.

______ **1.** A habitat is different from an ecosystem in that
- **a.** a habitat contains many ecosystems.
- **b.** an ecosystem can contain many habitats.
- **c.** a habitat is called a community.
- **d.** an ecosystem is the physical aspects of a habitat.

______ **2.** There may be billions of bacteria in
- **a.** a square kilometer of forest.
- **b.** an ecosystem.
- **c.** a handful of soil.
- **d.** the southeastern United States.

______ **3.** Which of the following is an example of an isolated ecosystem?
- **a.** a single rotting log on the forest floor
- **b.** a specific area of a lake
- **c.** an oceanic island
- **d.** All of the above

______ **4.** A regular progression of species replacement is called
- **a.** succession.
- **b.** ecology.
- **c.** forest development.
- **d.** a stable ecosystem.

______ **5.** Primary succession in a forest generally follows which sequence?
- **a.** forest fire→pioneer species→willows→hardwoods
- **b.** abandoned farm→shrubby weeds→cottonwoods→pine trees
- **c.** flood→mosses→ground cover→shrubs
- **d.** exposed rock→lichens→shrubs→trees→stable forest

In the space provided, write the letter of the description that best matches the term or phrase.

______ **6.** biotic factor

______ **7.** abiotic factor

______ **8.** mosses

______ **9.** hemlock trees

______ **10.** ecosystem

- **a.** a nonliving or physical aspect of a habitat
- **b.** typical stable forest organisms
- **c.** a community and all its physical aspects
- **d.** a living or biological aspect of a habitat
- **e.** typical pioneer organisms

Name ______________________ Class ____________ Date ____________

Assessment

Quiz

Section: Energy Flow in Ecosystems

In the space provided, write the letter of the term or phrase that best completes each statement or best answers each question.

______ **1.** Most of life on Earth depends on which of the following?
- **a.** animals that eat plants
- **b.** photosynthetic organisms
- **c.** animals that eat other animals
- **d.** consumers on the second trophic level

______ **2.** The lowest trophic level of any ecosystem is occupied by organisms such as
- **a.** lions, wolves, and snakes.
- **b.** humans, bears, and pigs.
- **c.** cows, horses, and caterpillars.
- **d.** plants, bacteria, and algae.

______ **3.** A tertiary consumer is one that is in
- **a.** the first trophic level.
- **b.** the second trophic level.
- **c.** the third trophic level.
- **d.** the fourth trophic level.

______ **4.** Food webs are formed from food chains because
- **a.** many individual animals feed at several trophic levels.
- **b.** energy flows better in several directions.
- **c.** all consumers depend on the same producers.
- **d.** herbivores can eat many types of plants.

______ **5.** The number of trophic levels that can be maintained in an ecosystem is limited by
- **a.** the number of species in the ecosystem.
- **b.** a gain in population size.
- **c.** the loss of potential energy.
- **d.** the number of individuals in the ecosystem.

In the space provided, write the letter of the description that best matches the term or phrase.

______ **6.** herbivore

______ **7.** carnivore

______ **8.** omnivore

______ **9.** detritivore

______ **10.** decomposer

- **a.** an animal that eats both plants and animals
- **b.** an animal that eats only primary producers
- **c.** an organism that causes decay
- **d.** an animal that eats only other animals
- **e.** an organism that eats organic waste and dead bodies

Name ______________________ Class ______________ Date ____________

Assessment

Quiz

Section: Cycling of Materials in Ecosystems

In the space provided, write the letter of the term or phrase that best completes each statement or best answers each question.

______ **1.** A biogeochemical cycle involves which aspect of the ecosystem?
- **a.** biological
- **b.** geological
- **c.** chemical
- **d.** All of the above

______ **2.** All of the following are cycled through living and nonliving parts of an ecosystem EXCEPT
- **a.** water.
- **b.** carbon.
- **c.** energy.
- **d.** phosphorus.

______ **3.** Which of the following contributes to the return of water vapor to the atmosphere?
- **a.** precipitation
- **b.** evaporation
- **c.** runoff
- **d.** percolation

______ **4.** Which process brings carbon into the living portion of its cycle?
- **a.** photosynthesis
- **b.** cellular respiration
- **c.** combustion
- **d.** decomposition

______ **5.** For which reason do organisms need phosphorus?
- **a.** Phosphorus is required by the roots of plants.
- **b.** It is part of the process of nitrogen fixation.
- **c.** Phosphorus is an essential part of ATP and DNA.
- **d.** It is bound to calcium as calcium phosphate, and calcium is required by organisms.

In the space provided, write the letter of the description that best matches the term or phrase.

______ **6.** respiration

______ **7.** ammonification

______ **8.** combustion

______ **9.** denitrification

______ **10.** erosion

- **a.** Carbon becomes available to other organisms through this process acting on limestone.
- **b.** This process returns nitrogen gas to the atmosphere.
- **c.** Carbon dioxide is a by-product of this biological reaction.
- **d.** This process releases carbon from fossil fuels.
- **e.** This process converts nitrogen into a useable form for organisms.

Name ______________________ Class ______________ Date ______________

Assessment

Chapter Test

Ecosystems

In the space provided, write the letter of the term or phrase that best completes each statement or best answers each question.

______ **1.** The production of ammonia by bacteria during the decay of nitrogen-containing urea is called
a. assimilation.
b. ammonification.
c. nitrification.
d. denitrification.

______ **2.** Grizzly bears, snakes, and worms can be members of the same
a. species.
b. trophic level.
c. ecosystem.
d. None of the above

______ **3.** All the organisms that live in a particular place and the physical aspects of that place make up a(n)
a. ecosystem.
b. habitat.
c. community.
d. food chain.

______ **4.** The number of species living in an ecosystem is referred to as
a. succession.
b. biodiversity.
c. the food chain.
d. productivity.

______ **5.** The most important abiotic factor for the organisms in an ecosystem is
a. climate.
b. sun.
c. weather.
d. water.

______ **6.** Animals that feed on plants are at least in which of the following?
a. first trophic level
b. second trophic level
c. third trophic level
d. fourth trophic level

______ **7.** The number of trophic levels in an ecological pyramid
a. is limitless.
b. is limited by the amount of energy that is lost at each trophic level.
c. never exceeds four.
d. never exceeds three.

______ **8.** The movement of substances, such as water and nitrogen, in a circular path between the nonliving environment and living organisms is called
a. a reservoir pathway.
b. photosynthesis.
c. a biogeochemical cycle.
d. succession.

______ **9.** In a typical primary succession initiated by a retreating glacier,
a. lichens and mosses precede trees.
b. the first plants are stunted by mineral deficiencies.
c. it takes about 10 years for trees to be able to thrive.
d. All of the above

Questions 10–12 refer to the figure below, which shows the feeding relationships in an Antarctic ecosystem.

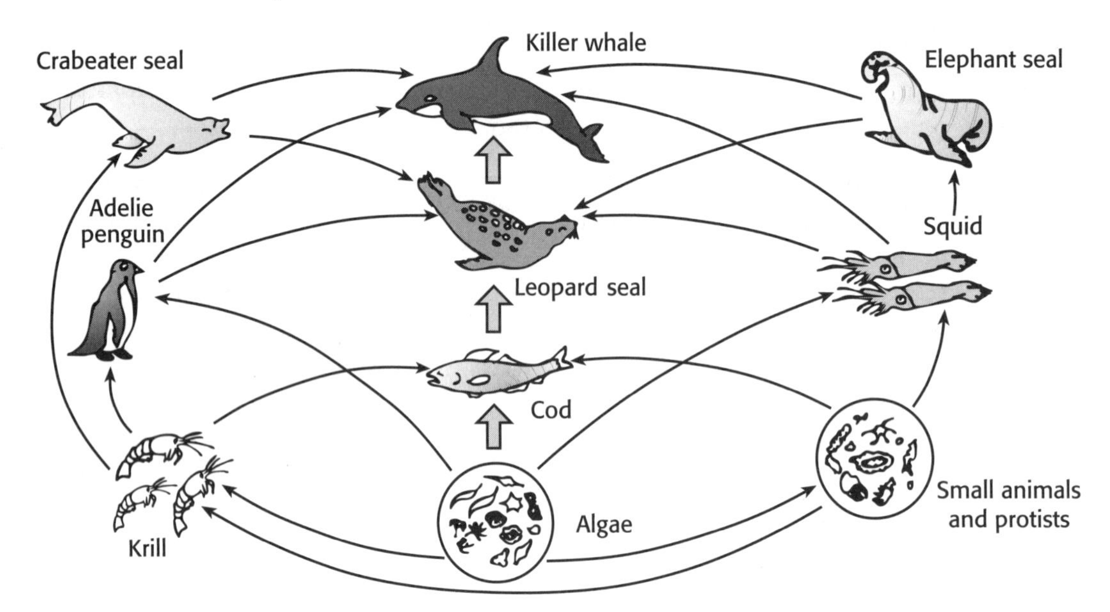

______ **10.** The figure above represents a

a. trophic net.
b. food chain.
c. food net.
d. food web.

______ **11.** The algae are

a. producers.
b. consumers.
c. parasites.
d. decomposers.

______ **12.** Leopard seals are

a. producers.
b. carnivores.
c. herbivores.
d. omnivores.

______ **13.** Organisms that obtain their energy from the organic wastes and dead bodies at all the energy levels in an ecosystem are called

a. decomposers.
b. detritivores.
c. consumers.
d. All of the above

______ **14.** Examples of elements that are recycled in an ecosystem include which of the following?

a. energy
b. water
c. carbon
d. ammonia

______**15.** Every time energy is transferred in an ecosystem, potential energy is lost
a. as heat.
b. due to weather.
c. because some animals die.
d. when it is recycled back to producers.

In the space provided, write the letter of the description that best matches the term or phrase.

______**16.** transpiration

______**17.** nitrogen fixation

______**18.** decomposers

______**19.** carbon dioxide

______**20.** biomass

a. release nutrients back to the environment to be recycled by other organisms

b. dry weight of tissue and organic matter in an ecosystem

c. a by-product of cellular respiration in nearly all living organisms

d. process by which water is returned to the atmosphere through plants

e. process by which nitrogen gas is converted to ammonia by bacteria

Name ______________________ Class ______________ Date ____________

Assessment

Chapter Test

Ecosystems

In the space provided, write the letter of the term or phrase that best completes each statement or best answers each question.

______ **1.** A typical ecosystem might include which of the following?
a. large and small mammals
b. microscopic eukaryotes
c. birds, trees, and flowers
d. All of the above

______ **2.** When succession occurs in areas where previous growth has occurred, it is called
a. secondary succession.
b. pioneer succession.
c. primary succession.
d. natural succession.

______ **3.** In 1866, the German biologist Ernst Haeckel named the study of how organisms fit in their environment, calling it
a. succession.
b. ecology.
c. environmental interaction.
d. population science.

______ **4.** In a meadow food chain, which is the correct sequence of the path of energy flow?
a. hawk → snake → mouse → grass
b. mouse → grass → hawk → snake
c. grass → mouse → snake → hawk
d. snake → mouse → hawk → grass

______ **5.** In a marine food web, there is far more total biomass in algae than in all the killer whales. Why is this so?
a. Whales are bigger than algae.
b. An alga has more mass than a killer whale.
c. Whales don't eat algae.
d. It takes a massive amount of algae to support a food web with a killer whale at the top.

______ **6.** The ultimate source of energy for producers and all consumers is
a. plants.
b. the sun.
c. algae.
d. the ocean.

______ **7.** Which of the following is the most accurately descriptive statement about the biodiversity in one square kilometer of pine forest in the southeastern United States?
a. Five kingdoms of life are represented.
b. The forest is home to many small mammals.
c. Birds inhabit the forest.
d. There are many different kinds of trees.

______ **8.** In a living portion of the water cycle, water passes through plants and evaporates into the atmosphere through the process of
a. photosynthesis.
b. respiration.
c. transpiration.
d. nitrification.

______ **9.** The carbon in the remains of organisms that lived long ago is released in the burning of
a. wood.
b. limestone.
c. calcium carbonate.
d. fossil fuels.

______ **10.** Nitrogen-fixing bacteria live in
a. the human intestine.
b. soil and plant roots.
c. rotting logs.
d. the atmosphere.

______ **11.** Humans cannot digest which of the following?
a. cellulose
b. carbohydrates
c. simple sugars
d. fats

______ **12.** Heat can be used to do mechanical work, but it is not useful in
a. trophic systems.
b. solar systems.
c. chemical systems.
d. biological systems.

______ **13.** Which material is required in the greatest quantity in all ecosystems?
a. manganese
b. sodium
c. water
d. iron

______ **14.** Marine organisms use carbon dioxide dissolved in seawater to
a. make calcium carbonate shells.
b. erode limestone.
c. grow bones and teeth.
d. engage in cellular respiration.

______ **15.** Which type of bacteria plays a significant role in the nitrogen cycle?
a. nitrogen-fixing bacteria
b. decomposers
c. denitrifying bacteria
d. All of the above

In the space provided, write the letter of the description that best matches the term or phrase.

______ **16.** habitat

______ **17.** biodiversity

______ **18.** primary productivity

______ **19.** food web

______ **20.** energy pyramid

______ **21.** water

______ **22.** nitrogen

a. complicated, interconnected group of food chains

b. a diagram that shows the flow of energy from the sun through all trophic levels

c. of abiotic factors, this has the greatest influence on an ecosystem's inhabitants

d. the rate at which organic material is produced by photosynthesis

e. cycled throughout the living world primarily by bacteria

f. the place where an organism or population of organisms lives

g. the number of species living in an ecosystem

Read each question, and write your answer in the space provided.

23. Why are both bacteria and fungi important organisms in an ecosystem?

____

____

____

24. Describe the process of primary succession that occurred following the retreat of the glacier at Glacier Bay, Alaska.

____

____

____

____

____

25. Why are producers an essential component of an ecosystem?

____

____

Name ______________________ Class ______________ Date ____________

Quick Lab

DATASHEET FOR IN-TEXT LAB

Evaluating Biodiversity

By making simple observations, you can draw some conclusions about biodiversity in an ecosystem.

MATERIALS

- note pad
- pencil

Procedure

1. **CAUTION: Do not approach or touch any wild animals. Do not disturb plants.** Prepare a list of biotic and abiotic factors that you observe around your home or in a nearby park.

Analysis

1. **Identify** the habitat and community that you observed.

____

2. **Calculate** the number of different species as a percentage of the total number of organisms that you saw.

____

____

3. **Rank** the importance of biotic factors within the ecosystem you observed.

____

____

____

4. **Infer** what the relationships are between biotic factors and abiotic factors in the observed ecosystem.

____

____

____

____

____

____

Name ______________________ Class ______________ Date ____________

Quick Lab

DATASHEET FOR IN-TEXT LAB

Modeling Succession

You can create a small ecosystem and measure how organisms modify their environment.

MATERIALS

- 1 qt glass jar with a lid
- one-half quart of pasteurized milk
- pH strips

Procedure

1. You will be using the data table below to record your data.

Data Table		
Day	**pH**	**Appearance**
1		
2		
3		
4		
5		
6		
7		

2. Half fill a quart jar with pasteurized milk, and cover the jar loosely with a lid. Measure and record the pH. Place the jar in a 37°C incubator.
3. Check and record the pH of the milk with pH strips every day for seven days. As milk spoils, its pH changes. Different populations of microorganisms become established, alter substances in the milk, and then die off when conditions no longer favor their survival.
4. Record any visible changes in the milk each day.

Analysis

1. **Identify** what happened to the pH of the milk as time passed.

____

____

____

____

2. **Infer** what the change in pH means about the populations of microorganisms in the milk.

____

____

____

____

____

3. **Critical Thinking**
   Evaluating Results How does this model confirm the model of succession in Glacier Bay?

____

____

____

____

____

Name ______________________ Class ______________ Date ____________

Exploration Lab

DATASHEET FOR IN-TEXT LAB

Modeling Ecosystem Change over Time

SKILLS

- Using scientific methods
- Modeling
- Observing

OBJECTIVES

- **Construct** a model ecosystem.
- **Observe** the interactions of organisms in a model ecosystem.
- **Predict** how the number of each species in a model ecosystem will change over time.
- **Compare** a model ecosystem with a natural ecosystem.

MATERIALS

- coarse sand or pea gravel
- large glass jar with a lid or terrarium
- soil
- pinch of grass seeds
- pinch of clover seeds
- mung bean seeds
- earthworms
- isopods (pill bugs)
- mealworms (beetle larva)
- crickets

Before You Begin

Organisms in an **ecosystem** interact with each other and with their environment. One of the interactions that occurs among the organisms in an ecosystem is feeding. A **food web** describes the feeding relationships among the organisms in an ecosystem. In this lab, you will model a natural ecosystem by building a **closed ecosystem** in a bottle or a jar. You will then observe the interactions of the organisms in the ecosystem and note any changes that occur over time.

1. Write a definition for each boldface term in the paragraph above and for each of the following terms: producer, decomposer, consumer, herbivore, carnivore, trophic level. Use a separate sheet of paper.

2. Based on the objectives for this lab, write a question you would like to explore about ecosystems.

____

____

Procedure

PART A: BUILDING AN ECOSYSTEM IN A JAR

1. Place 2 in. of sand or pea gravel in the bottom of a large, clean glass jar with a lid. **CAUTION: Glassware is fragile. Notify your teacher promptly of any broken glass or cuts. Do not clean up broken glass or spills with broken glass unless your teacher tells you to do so.** Cover the gravel with 2 in. of soil.

2. Sprinkle the seeds of two or three types of small plants, such as grasses and clovers, on the surface of the soil. Put a lid on the jar, and place it in indirect sunlight. Let the jar remain undisturbed for a week.

3. After one week, place the animals into the jar and replace the lid. Place the lid on the jar loosely to enable air entry for the animals. You may also put small holes in the lid.

You Choose

As you design your experiment, decide the following:

- **a.** what question you will explore
- **b.** what hypothesis you will test
- **c.** how you will plant the seeds
- **d.** where you will place the ecosystem for one week so that it remains undisturbed and in indirect sunlight
- **e.** how often you will add water to the ecosystem after the first week
- **f.** how many of each organism you will use
- **g.** what data you will record in your data table

PART B: DESIGN AN EXPERIMENT

4. Work with the members of your lab group to explore one of the questions written for step 2 of **Before You Begin.** To explore the question, design an experiment that uses the materials listed for this lab.

5. Write a procedure for your experiment. Make a list of all the safety precautions you will take. Have your teacher approve your procedure and safety precautions before you begin the experiment.

6. Set up your group's experiment. Conduct your experiment for at least 14 days.

PART C: CLEANUP AND DISPOSAL

7. Dispose of solutions, broken glass, and other materials in the designated waste containers. Do not put lab materials in the trash unless your teacher tells you to do so.

8. Clean up your work area and all lab equipment. Return lab equipment to its proper place. Wash your hands thoroughly before you leave the lab and after you finish all work.

Analyze and Conclude

1. **Summarizing Results** Make graphs showing how the number of individuals of each species in your ecosystem changed over time. Plot time on the x-axis and the number of organisms on the y-axis.

____

____

____

2. **Analyzing Results** How did your results compare with your hypothesis? Explain any differences.

____

____

____

3. **Inferring Conclusions** Construct a food web for the ecosystem you observed.

____

____

____

____

4. **Recognizing Relationships** Does your model ecosystem resemble a natural ecosystem? Explain.

____

____

____

____

____

5. Analyzing Methods How might you have built your model ecosystem differently to better represent a natural ecosystem?

____

____

____

____

6. Evaluating Methods Was your model ecosystem truly a "closed ecosystem"? List your model's strengths and weaknesses as a closed ecosystem.

____

____

____

____

____

____

7. Further Inquiry Write a new question about ecosystems that you could explore with another investigation.

____

____

____

Name ______________________ Class ______________ Date ____________

Quick Lab

MODELING

Making a Food Web

One organism consumes another for energy and raw materials. A food chain shows the sequence in which energy passes from one organism to another as it flows through a community. In this lab, you will draw food chains that might be found in a woodland community and show how the food chains are connected to form a food web.

OBJECTIVES

Depict three food chains within a woodland community.

Combine the food chains into a food web.

MATERIALS

- paper
- pen or pencil

Procedure

1. Closely observe **Figure 1,** which shows a portion of a woodland community. List all the organisms that you see.

____

____

____

FIGURE 1 A WOODLAND COMMUNITY

2. Add to your list other organisms that might be present in this community but are not shown.

____

____

3. On a separate sheet of paper, write the name of one organism from your list that is capable of photosynthesis.
4. Draw a short arrow leading from this organism to the name of a second organism that might eat it. These are the first two links of a food chain.
5. Extend your chain to three links by adding an arrow and a third organism that might consume the second.
6. Extend your food chain to four links.
7. Make two more food chains consisting of four links each.
8. Construct a food web by drawing arrows to show how organisms in the three food chains are linked to one another. Make as many connections as possible.

Analysis and Conclusions

1. **Analyzing Data** How are food chains and food webs alike? How are they different?

____

____

____

____

2. **Evaluating Models** How is a diagram of a food web more helpful than a written description of the same information?

____

____

3. **Drawing Conclusions** If all the green plants were removed from the woodland community, how might the flow of energy be affected? Explain your answer.

____

____

4. **Drawing Conclusions** If the top-level consumers were eliminated from a food web, would the populations in the levels below them increase without bounds? Why or why not?

____

____

Name ______________________ Class ______________ Date ____________

Skills Practice Lab

CBL™ PROBEWARE

Assessing Abiotic Factors in the Environment

Every ecosystem contains specific biological and physical components. The biological components are the organisms that live in the ecosystem. The physical components of an ecosystem are called *abiotic factors*. In terrestrial ecosystems, abiotic factors include light intensity, relative humidity, air temperature, soil temperature, soil moisture, and soil pH.

Abiotic factors determine which organisms can exist in a particular environment. Plants that require bright sunlight, for example, will not survive in the deep shade of a forest floor. Abiotic factors also influence the rate of growth, reproductive potential, and genetic expression of organisms. In acidic soil, for example, hydrangeas produce modified leaves, called bracts, that are blue. In alkaline soil, the bracts are pink.

In this lab, you will measure six abiotic factors in the environment at one location. You will then compare your findings with those from other locations.

OBJECTIVES

Test light intensity, relative humidity, air temperature, soil temperature, and soil pH in an environment.

Calculate the percentage of soil moisture in that environment.

Compare the tested abiotic factors with those of other environments.

MATERIALS

- balance
- beaker, 100 mL
- beaker, 500 mL
- CBL System
- filter
- filter papers (2)
- garden trowel
- graduated cylinder, 100 mL
- jar with lid
- lab apron
- light probe
- metal can with both ends removed
- metal spatula
- meterstick
- mortar and pestle
- nail, large
- oven mitts
- paper plate
- permanent marker
- pH probe
- relative humidity sensor
- rinse bottle filled with deionized water
- rubber band
- safety goggles
- sandwich bag, sealable plastic
- sieve or strainer
- temperature probe
- TI graphing calculator

Procedure

DAY 1: SETTING UP THE CBL SYSTEM

1. Plug the temperature probe into the Channel 1 input of the CBL unit. Connect the relative humidity sensor to the Channel 2 input. Plug the light probe into the Channel 3 input. Use the black cable to connect the CBL unit to the graphing calculator.

2. Use a rubber band to fasten the temperature probe, relative humidity sensor, and light probe together.

3. Turn on both the CBL unit and the calculator. Start the CHEMBIO program and go to the MAIN MENU.

4. Select SET UP PROBES. Enter "3" as the number of probes. Select TEMPERATURE from the SELECT PROBE menu. Enter "1" as the channel number.

5. Select MORE PROBES from the SELECT PROBE menu. Select REL HUMID from the SELECT PROBE menu. Enter "2" as the channel number. Select USE STORED from the CALIBRATION menu.

6. Select MORE PROBES from the SELECT PROBE menu. Select LIGHT from the SELECT PROBE menu. Enter "3" as the channel number.

DAY 1: MEASURING AIR TEMPERATURE, RELATIVE HUMIDITY, AND LIGHT INTENSITY

7. Go to the location where you will make your measurements. Find a spot of bare soil. If the ground is covered with fallen leaves, brush them aside to expose the soil.

8. Hold a meterstick vertically with the 0 cm end touching the soil. Hold the three probes at the top of the meterstick. The light probe should be pointing straight down.

9. Select COLLECT DATA from the MAIN MENU. Select MONITOR INPUT from the DATA COLLECTION menu.

10. Wait for the readings of all three probes to stabilize. Then record their values in **Table 1.**

11. Move the probes down 10 cm to the 90 cm mark. Repeat step 10.

12. Continue recording the temperature, relative humidity, and light intensity at 10 cm intervals. For the last recording, hold the sensors as close to the 0 cm mark as you can without letting them touch the soil. Record all values in **Table 1.**

TABLE 1 AIR TEMPERATURE, RELATIVE HUMIDITY, AND LIGHT INTENSITY

Height (cm)	Air temperature (°C)	Relative humidity (%)	Light intensity ($\mu W/cm^2$)
100			
90			
80			
70			
60			
50			
40			
30			
20			
10			
0			

DAY 1: MEASURING SOIL TEMPERATURE AND COLLECTING A SOIL SAMPLE

13. Push a large nail into the soil to a depth of 10 cm. Remove the nail.

14. Loosen the rubber band that holds the probes together, and insert the temperature probe all the way into the hole. When the temperature reading stabilizes, record the soil temperature in **Table 2.**

TABLE 2 SOIL TEMPERATURE, MOISTURE, AND pH

Soil temperature (°C)	
Mass of soil before drying (g)	
Mass of dried soil (g)	
Mass of soil moisture (g)	
Percentage of soil moisture	
Soil pH	

15. Turn off the CBL unit and the calculator.

16. Use a garden trowel to dig into the soil to a depth of 10 cm. Collect enough soil from that depth to fill a plastic sandwich bag about half full. Seal the bag, and return to the lab with the soil sample and the rest of your materials. Wash your hands thoroughly.

DAY 1: MEASURING THE MASS OF THE SOIL SAMPLE

FIGURE 1 SOIL CONTAINER

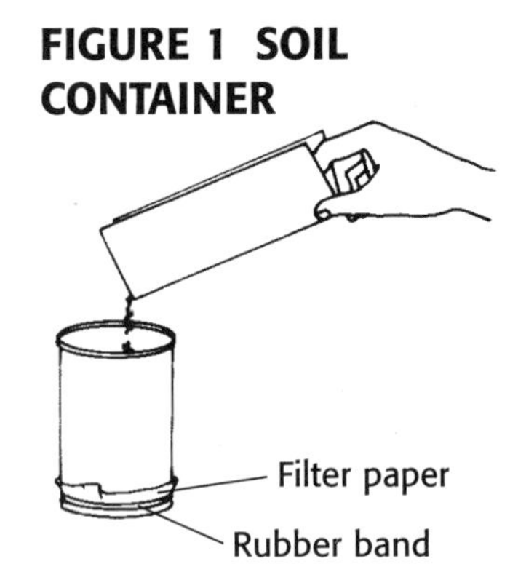

17. Place a sheet of filter paper over one end of a metal can with both ends removed, as shown in **Figure 1.** Secure the filter paper with a rubber band.

18. Use a balance to measure the mass of the soil you collected. Record the mass in **Table 2.**

19. Place the soil in the container you made in step 17. Use a permanent marker to label your container with the initials of each member of your group. Place your container in a drying oven set at 100°C for 24 hours.

DAY 2: MEASURING THE MASS OF THE DRIED SOIL SAMPLE

20. Put on safety goggles and a lab apron.

21. Using oven mitts, carefully remove your container from the oven. **CAUTION: Do not touch the container with your bare hands.** Empty the dried soil from the container onto a paper plate. Use a metal spatula to dislodge any soil that sticks to the container.

22. Use a balance to measure the mass of the dried soil. Record the mass in **Table 2.**

DAY 2: PREPARING THE SOIL SAMPLE FOR pH MEASUREMENT

23. Use a sieve or strainer to remove large particles such as rocks from the dried soil. Grind 30 g of the sieved soil with a mortar and pestle.

24. Place the ground soil in a jar. **CAUTION: If a piece of glassware breaks, do not pick it up with your bare hands. Notify your teacher immediately.** Add 60 mL of deionized water to the jar. Attach the lid and shake the jar vigorously for 1 minute.

25. Filter the soil-water mixture, collecting the filtrate in a 100 mL beaker.

DAY 2: SETTING UP THE CBL SYSTEM

26. Connect the pH probe into the Channel 1 input of the CBL unit. Use the black cable to connect the CBL unit to the graphing calculator.

27. Turn on both the CBL unit and the calculator. Start the CHEMBIO program and go to the MAIN MENU.

28. Select SET UP PROBES. Enter "1" as the number of probes. Select pH from the SELECT PROBE menu. Enter "1" as the channel number.

29. Select USE STORED from the CALIBRATION menu.

30. Select COLLECT DATA from the MAIN MENU. Select MONITOR INPUT from the DATA COLLECTION menu.

DAY 2: MEASURING THE pH OF THE SOIL-WATER FILTRATE

31. Remove the pH probe from its storage solution. Carefully rinse the probe with deionized water, catching the rinse water in a 500 mL beaker.

32. Submerge the pH probe into the filtrate. When the pH reading stabilizes, record it in **Table 2.**

33. Turn off the CBL unit and the calculator. Rinse the pH probe thoroughly with deionized water, and return the probe to its storage solution, making sure that the cap is on tight. Dispose of the filtrate and the rinse water as instructed by your teacher.

34. Share the data you have recorded in **Table 1** and **Table 2** with the members of the other groups in your class.

35. Put away your materials, and clean up your work area. Wash your hands thoroughly before leaving the lab.

Analysis

1. Constructing Graphs On the graph in **Figure 2,** plot the air temperature versus the height above ground at the location you studied. Label the y-axis.

FIGURE 2 AIR TEMPERATURE VERSUS HEIGHT

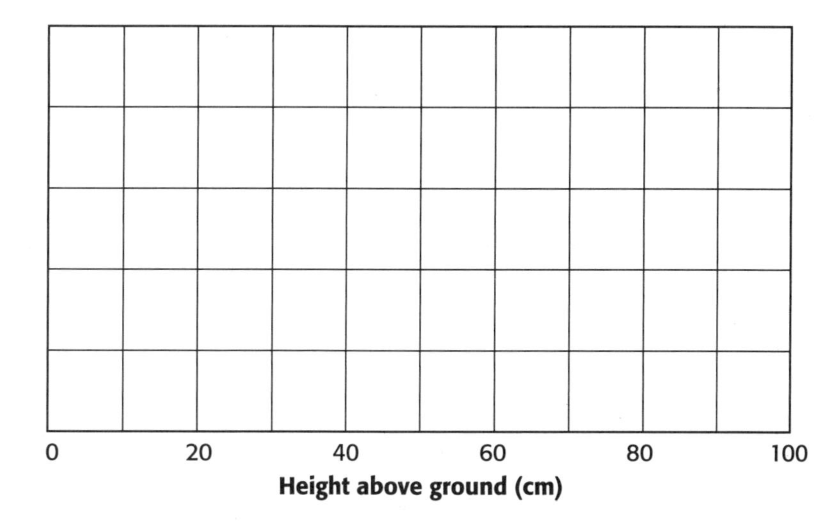

2. Constructing Graphs On the graph in **Figure 3,** plot the relative humidity versus the height above ground at the location you studied.

FIGURE 3 RELATIVE HUMIDITY VERSUS HEIGHT

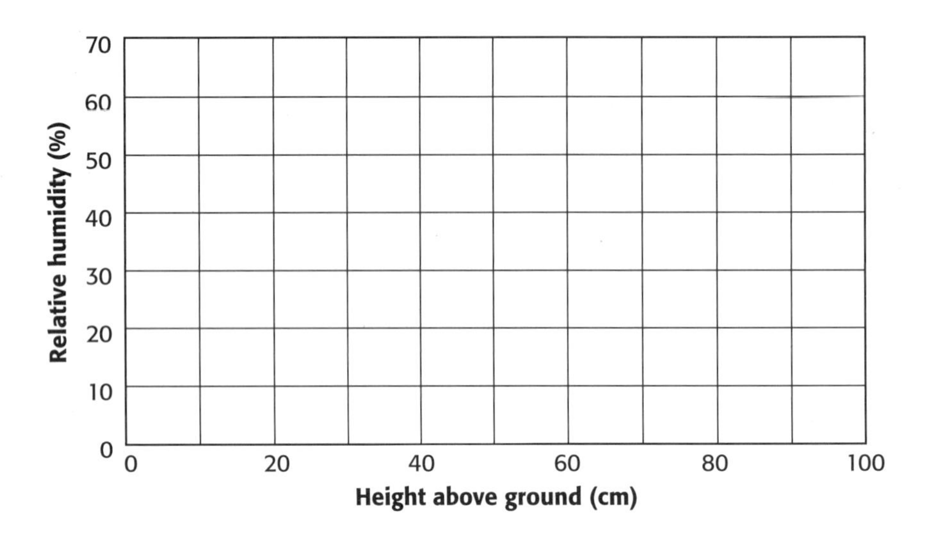

3. Constructing Graphs On the graph in **Figure 4,** plot the light intensity versus the height above ground at the location you studied.

FIGURE 4 LIGHT INTENSITY VERSUS HEIGHT

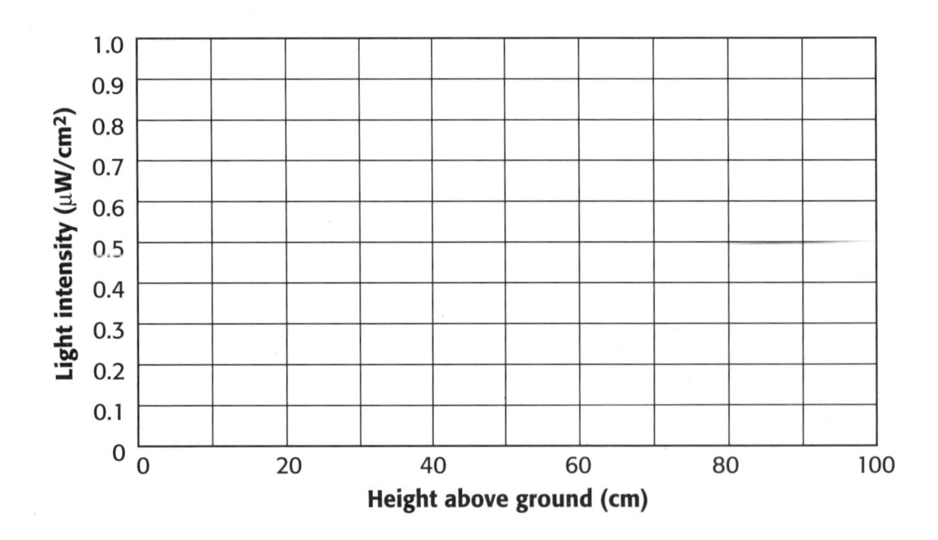

4. Organizing Data Calculate the mass of soil moisture in your sample by subtracting the mass of dried soil from the mass of soil before drying. Record the result in **Table 2.**

5. Organizing Data Calculate the percentage of soil moisture in your sample, using the following formula: percentage of soil moisture = (mass of soil moisture ÷ mass of soil before drying) × 100. Record the result in **Table 2.**

Conclusions

1. **Analyzing Graphs** How did the air temperature, relative humidity, and light intensity vary with the height above ground at the location you studied?

2. **Evaluating Methods** Why is it important to measure abiotic factors, such as air temperature, relative humidity, and light intensity, at different heights?

3. **Evaluating Results** Compare the air temperature, relative humidity, and light intensity at the location you studied with those abiotic factors at the location studied by another group.

4. **Interpreting Information** How did the soil temperature at a depth of 10 cm compare with the air temperature at 0 cm? Account for any difference.

5. Drawing Conclusions Compare the percentage of soil moisture at the location you studied with that at the location studied by another group. Provide a reasonable explanation for any difference.

____

____

____

____

6. Making Predictions If you were to plant a hydrangea at the location you studied, what would be the color of its bracts? Explain your reasoning.

____

____

____

Extensions

1. **Research and Communications** *Soil scientists* study all aspects of the soil, from its origin to its composition and distribution. A soil scientist may test the effects of fertilizer on soil quality, assess how a field drains, or develop practices to increase soil productivity. Conducting chemical analysis of soil and examining soil for bacterial and mineral content are other jobs that soil scientists perform. Find out what training and skills are needed to become a soil scientist. Report your findings in a short paper or oral presentation.
2. **Research and Communications** Select a plant and determine what it needs to thrive in any given region. List the abiotic factors that influence the survival of the plant. Present your list in a poster.
3. **Designing Experiments** Other important abiotic factors in terrestrial environments include soil nutrients and wind. Devise a procedure to measure these factors in different locations.

Name ______________________ Class ______________ Date ______________

Quick Lab

DATASHEET FOR IN-TEXT LAB

Evaluating Biodiversity

By making simple observations, you can draw some conclusions about biodiversity in an ecosystem.

MATERIALS

- note pad
- pencil

Procedure

1. **CAUTION: Do not approach or touch any wild animals. Do not disturb plants.** Prepare a list of biotic and abiotic factors that you observe around your home or in a nearby park.

Analysis

1. **Identify** the habitat and community that you observed.

   Answers will vary.

2. **Calculate** the number of different species as a percentage of the total number of organisms that you saw.

   Answers will vary. Example: 3 robins/29 total organisms = 0.103 = 10%

3. **Rank** the importance of biotic factors within the ecosystem you observed.

   Answers will vary.

4. **Infer** what the relationships are between biotic factors and abiotic factors in the observed ecosystem.

   Answers will vary. In general, the abiotic factors in an ecosystem provide organisms (biotic factors) with a physical place to live, energy, nutrients, and water. The organisms alter and recycle some of these abiotic factors, changing the landscape in the process.

Name ______________________ Class ______________ Date ______________

Quick Lab

DATASHEET FOR IN-TEXT LAB

Modeling Succession

You can create a small ecosystem and measure how organisms modify their environment.

MATERIALS

- 1 qt glass jar with a lid
- one-half quart of pasteurized milk
- pH strips

Procedure

1. You will be using the data table provided to record your data.

Data Table		
Day	**pH**	**Appearance**
1		
2		
3		
4		
5		
6		
7		

2. Half fill a quart jar with pasteurized milk, and cover the jar loosely with a lid. Measure and record the pH. Place the jar in a 37°C incubator.
3. Check and record the pH of the milk with pH strips every day for seven days. As milk spoils, its pH changes. Different populations of microorganisms become established, alter substances in the milk, and then die off when conditions no longer favor their survival.
4. Record any visible changes in the milk each day.

Name ______________________ Class ______________ Date ____________

Analysis

1. **Identify** what happened to the pH of the milk as time passed.

   The pH dropped as the environment became more acidic.

2. **Infer** what the change in pH means about the populations of microorganisms in the milk.

   By-products from the microorganisms change the pH. Then new organisms that are better adapted to the changed pH begin to thrive.

3. **Critical Thinking**
   Evaluating Results How does this model confirm the model of succession in Glacier Bay?

   Like the Glacier Bay model, organisms colonize and slowly change a new environment such that it becomes more suitable for other organisms.

Name ______________________ Class ______________ Date ____________

Exploration Lab

DATASHEET FOR IN-TEXT LAB

Modeling Ecosystem Change over Time

SKILLS

- Using scientific methods
- Modeling
- Observing

OBJECTIVES

- **Construct** a model ecosystem.
- **Observe** the interactions of organisms in a model ecosystem.
- **Predict** how the number of each species in a model ecosystem will change over time.
- **Compare** a model ecosystem with a natural ecosystem.

MATERIALS

- coarse sand or pea gravel
- large glass jar with a lid or terrarium
- soil
- pinch of grass seeds
- pinch of clover seeds
- mung bean seeds
- earthworms
- isopods (pill bugs)
- mealworms (beetle larva)
- crickets

Before You Begin

Organisms in an **ecosystem** interact with each other and with their environment. One of the interactions that occurs among the organisms in an ecosystem is feeding. A **food web** describes the feeding relationships among the organisms in an ecosystem. In this lab, you will model a natural ecosystem by building a **closed ecosystem** in a bottle or a jar. You will then observe the interactions of the organisms in the ecosystem and note any changes that occur over time.

1. Write a definition for each boldface term in the paragraph above and for each of the following terms: producer, decomposer, consumer, herbivore, carnivore, trophic level. Use a separate sheet of paper. **Answers appear in the TE for this lab.**

2. Based on the objectives for this lab, write a question you would like to explore about ecosystems.

Answers will vary. For example, what are the effects of continuous exposure to bright light on the ecosystem?

Procedure

PART A: BUILDING AN ECOSYSTEM IN A JAR

1. Place 2 in. of sand or pea gravel in the bottom of a large, clean glass jar with a lid. **CAUTION: Glassware is fragile. Notify your teacher promptly of any broken glass or cuts. Do not clean up broken glass or spills with broken glass unless your teacher tells you to do so.** Cover the gravel with 2 in. of soil.

2. Sprinkle the seeds of two or three types of small plants, such as grasses and clovers, on the surface of the soil. Put a lid on the jar, and place it in indirect sunlight. Let the jar remain undisturbed for a week.

3. After one week, place the animals into the jar and replace the lid. Place the lid on the jar loosely to enable air entry for the animals. You may also put small holes in the lid.

You Choose

As you design your experiment, decide the following:

a. what question you will explore

b. what hypothesis you will test

c. how you will plant the seeds

d. where you will place the ecosystem for one week so that it remains undisturbed and in indirect sunlight

e. how often you will add water to the ecosystem after the first week

f. how many of each organism you will use

g. what data you will record in your data table

PART B: DESIGN AN EXPERIMENT

4. Work with the members of your lab group to explore one of the questions written for step 2 of **Before You Begin.** To explore the question, design an experiment that uses the materials listed for this lab.

5. Write a procedure for your experiment. Make a list of all the safety precautions you will take. Have your teacher approve your procedure and safety precautions before you begin the experiment.

6. Set up your group's experiment. Conduct your experiment for at least 14 days.

Name ______________________ Class ______________ Date ______________

PART C: CLEANUP AND DISPOSAL

7. Dispose of solutions, broken glass, and other materials in the designated waste containers. Do not put lab materials in the trash unless your teacher tells you to do so.

8. Clean up your work area and all lab equipment. Return lab equipment to its proper place. Wash your hands thoroughly before you leave the lab and after you finish all work.

Analyze and Conclude

1. Summarizing Results Make graphs showing how the number of individuals of each species in your ecosystem changed over time. Plot time on the x-axis and the number of organisms on the y-axis.

Answers will vary. Students should make one graph for each species observed or use different colors to indicate each species.

2. Analyzing Results How did your results compare with your hypothesis? Explain any differences.

Answers will vary.

3. Inferring Conclusions Construct a food web for the ecosystem you observed.

Answer will vary. All plants are producers (primary trophic level); earthworms feed on dead plant material in the soil; crickets feed on plants; mealworms (beetle larvae) feed on plants; isopods (pill bugs) eat wood.

4. Recognizing Relationships Does your model ecosystem resemble a natural ecosystem? Explain.

Yes and no. Natural ecosystems and the model ecosystem both contain organisms at several trophic levels, have living and nonliving components, and depend on the sun for energy. However, the model ecosystem is less diverse, much younger, and has more definite boundaries than a natural ecosystem.

5. **Analyzing Methods** How might you have built your model ecosystem differently to better represent a natural ecosystem?

   Answers will vary.

6. **Evaluating Methods** Was your model ecosystem truly a "closed ecosystem"? List your model's strengths and weaknesses as a closed ecosystem.

   No, strengths are that the organisms in the model ecosystem did not leave the ecosystem and that other organisms could not enter from the outside. Weaknesses are that water and air probably had to be added to maintain a healthy ecosystem.

7. **Further Inquiry** Write a new question about ecosystems that you could explore with another investigation.

   Answers will vary. For example: What are the effects of certain abiotic factors, such as temperature, light, and moisture, on the organisms in an ecosystem?

Quick Lab

MODELING

Making a Food Web

Teacher Notes

TIME REQUIRED 20 minutes

SKILLS ACQUIRED

Communicating
Constructing models
Inferring
Organizing and analyzing data

RATINGS Easy ← 1 2 3 4 → Hard

Teacher Prep–1
Student Setup–1
Concept Level–2
Cleanup–1

THE SCIENTIFIC METHOD

Make Observations Step 1 of the Procedure requires students to make observations.

Analyze the Results Analysis and Conclusions question 1 asks students to analyze their results.

Draw Conclusions Analysis and Conclusions questions 3 and 4 ask students to draw conclusions based on their results.

TIPS AND TRICKS

This lab works best in groups of two or individually.

No outside preparation is required for this lab, but showing slides, transparencies, or photographs of organisms common to a woodland habitat can be helpful. To enhance the lab, you may wish to provide nature magazines and similar materials from which students can cut out pictures of the organisms in their food chains.

You may want to substitute a local community for the woodland community shown or use a community that is more familiar to your students. Discuss the types of organisms that might inhabit the community.

After students make their food web, have them follow the individual food chains within the web and determine the total number of chains.

Name ______________________ Class ______________ Date ______________

Quick Lab

MODELING

Making a Food Web

One organism consumes another for energy and raw materials. A food chain shows the sequence in which energy passes from one organism to another as it flows through a community. In this lab, you will draw food chains that might be found in a woodland community and show how the food chains are connected to form a food web.

OBJECTIVES

Depict three food chains within a woodland community.

Combine the food chains into a food web.

MATERIALS

- paper
- pen or pencil

Procedure

1. Closely observe **Figure 1,** which shows a portion of a woodland community. List all the organisms that you see.

   Students should list all the organisms visible in the illustration.

FIGURE 1 A WOODLAND COMMUNITY

Name ______________________ Class ______________ Date ____________

Making a Food Web *continued*

2. Add to your list other organisms that might be present in this community but are not shown.

   Students might list microorganisms and other fungi, plants, invertebrates, and vertebrates.

3. On a separate sheet of paper, write the name of one organism from your list that is capable of photosynthesis.
4. Draw a short arrow leading from this organism to the name of a second organism that might eat it. These are the first two links of a food chain.
5. Extend your chain to three links by adding an arrow and a third organism that might consume the second.
6. Extend your food chain to four links.
7. Make two more food chains consisting of four links each.
8. Construct a food web by drawing arrows to show how organisms in the three food chains are linked to one another. Make as many connections as possible.

Analysis and Conclusions

1. **Analyzing Data** How are food chains and food webs alike? How are they different?

   Both food chains and food webs show the flow of energy from organism to organism within a community. A food web is more complex than a food chain because it shows how the individual food chains within the community are linked to one another.

2. **Evaluating Models** How is a diagram of a food web more helpful than a written description of the same information?

   Because a food web diagram is a graphical representation, it can be used to see more easily how energy flows through a particular community.

3. **Drawing Conclusions** If all the green plants were removed from the woodland community, how might the flow of energy be affected? Explain your answer.

   The flow of energy would stop because green plants, which use the sun's energy to make food, are the base of the community's available energy.

4. **Drawing Conclusions** If the top-level consumers were eliminated from a food web, would the populations in the levels below them increase without bounds? Why or why not?

   No, population size is affected by many factors, including predation, weather, available food, available space, disease, and other biotic and abiotic factors.

Skills Practice Lab

CBL™ PROBEWARE

Assessing Abiotic Factors in the Environment

Teacher Notes

TIME REQUIRED Two 45-minute periods

SKILLS ACQUIRED

Collecting data
Experimenting
Identifying patterns
Interpreting
Measuring
Organizing and analyzing data
Predicting

RATINGS

Easy ← 1 2 3 4 → Hard

Teacher Prep–2
Student Setup–3
Concept Level–2
Cleanup–1

THE SCIENTIFIC METHOD

Make Observations Students observe abiotic factors in a terrestrial environment.

Analyze the Results Analysis questions 4 and 5 and Conclusions questions 1, 3, and 4 ask students to analyze their graphs.

Draw Conclusions Conclusions question 5 asks students to draw conclusions about data from different locations.

MATERIALS

Materials for this lab can be purchased from WARD'S. See the *Master Materials List* for ordering instructions.

Instead of using a drying oven, students can dry the soil by breaking it up and placing it on a paper plate left under a lamp. This drying process, however, will take at least 2 days.

SAFETY CAUTIONS

- Discuss all safety symbols and caution statements with students.
- Before allowing students to go into the field, survey the sites for potential dangers. Show students pictures of any dangerous organisms they might encounter, such as poison ivy, poison oak, black widow spiders, or poisonous snakes. Discuss how students can avoid problems with those organisms.
- Instruct students not to wear sandals or open-toed shoes in the field.

DISPOSAL

- Have students return the soil samples to the locations where they were collected.
- The soil filtrate and pH probe rinse water may be poured down the drain with large amounts of water.

TIPS AND TRICKS

Preparation

This lab works best in groups of two to four students.

Select locations around the school where the abiotic factors students are to study are likely to be different. Such locations may include shaded sites and sites exposed to direct sunlight, sites with bare soil or vegetative ground cover, and sites with different types of soil.

Prepare a map showing the locations where students are to make their outdoor measurements and collect soil samples. The number of locations will depend on class size and number of students per group.

This lab's procedure calls for using the stored calibration values for the pH probes. Recalibrating the pH probes might give more accurate results. To recalibrate, you can use two standard buffer solutions of pH 4 and pH 7. Refer to the pH probe booklet for instructions on calibrating probes.

You can either purchase pH 7 buffer solution or make it by combining 500 mL of 0.10 M Na_2PO_4 with 291 mL of 0.10 M NaOH and diluting to 1 L with deionized water. Check the pH and adjust to 7 if necessary.

The procedure in this lab is written for use with the original CBL system. If you are using CBL 2 or LabPro, the CHEMBIO program can still be used. Updated versions of this program can be downloaded from **www.vernier.com.** For additional information on how to integrate the CBL system into your laboratory, see the Program Introduction.

Some sensors may require the use of an adapter. Students will need to connect the adapter to the sensor before connecting it to the CBL.

Procedure

The nail that students use in step 13 should be at least 10 cm long and about the same diameter as the temperature probe. Caution students not to force the temperature probe into the soil.

Direct students to leave each study location as they found it and to fill in the holes they made when they return the soil.

Tell students not to let their pH probe dry out. Remind them not to hit the sides or bottom of the beaker with the probe when they test the pH of the filtrate.

For Figure 2, students should label the y-axis with a range of temperatures that are appropriate for the day of the lab.

Students who wish to complete Extension item 3 can use an anemometer to measure wind speed and a soil-test kit to measure the concentration of soil nutrients, such as nitrate, phosphate, and potassium. These items can be purchased from WARD'S. Measurements of wind speed should be made at different heights because wind speed is lower near the surface of the ground.

Name ______________________ Class ______________ Date ____________

Skills Practice Lab

CBL™ PROBEWARE

Assessing Abiotic Factors in the Environment

Every ecosystem contains specific biological and physical components. The biological components are the organisms that live in the ecosystem. The physical components of an ecosystem are called *abiotic factors*. In terrestrial ecosystems, abiotic factors include light intensity, relative humidity, air temperature, soil temperature, soil moisture, and soil pH.

Abiotic factors determine which organisms can exist in a particular environment. Plants that require bright sunlight, for example, will not survive in the deep shade of a forest floor. Abiotic factors also influence the rate of growth, reproductive potential, and genetic expression of organisms. In acidic soil, for example, hydrangeas produce modified leaves, called bracts, that are blue. In alkaline soil, the bracts are pink.

In this lab, you will measure six abiotic factors in the environment at one location. You will then compare your findings with those from other locations.

OBJECTIVES

Test light intensity, relative humidity, air temperature, soil temperature, and soil pH in an environment.

Calculate the percentage of soil moisture in that environment.

Compare the tested abiotic factors with those of other environments.

MATERIALS

- balance
- beaker, 100 mL
- beaker, 500 mL
- CBL System
- filter
- filter papers (2)
- garden trowel
- graduated cylinder, 100 mL
- jar with lid
- lab apron
- light probe
- metal can with both ends removed
- metal spatula
- meterstick
- mortar and pestle
- nail, large
- oven mitts
- paper plate
- permanent marker
- pH probe
- relative humidity sensor
- rinse bottle filled with deionized water
- rubber band
- safety goggles
- sandwich bag, sealable plastic
- sieve or strainer
- temperature probe
- TI graphing calculator

Procedure

DAY 1: SETTING UP THE CBL SYSTEM

1. Plug the temperature probe into the Channel 1 input of the CBL unit. Connect the relative humidity sensor to the Channel 2 input. Plug the light probe into the Channel 3 input. Use the black cable to connect the CBL unit to the graphing calculator.
2. Use a rubber band to fasten the temperature probe, relative humidity sensor, and light probe together.
3. Turn on both the CBL unit and the calculator. Start the CHEMBIO program and go to the MAIN MENU.
4. Select SET UP PROBES. Enter "3" as the number of probes. Select TEMPERATURE from the SELECT PROBE menu. Enter "1" as the channel number.
5. Select MORE PROBES from the SELECT PROBE menu. Select REL HUMID from the SELECT PROBE menu. Enter "2" as the channel number. Select USE STORED from the CALIBRATION menu.
6. Select MORE PROBES from the SELECT PROBE menu. Select LIGHT from the SELECT PROBE menu. Enter "3" as the channel number.

DAY 1: MEASURING AIR TEMPERATURE, RELATIVE HUMIDITY, AND LIGHT INTENSITY

7. Go to the location where you will make your measurements. Find a spot of bare soil. If the ground is covered with fallen leaves, brush them aside to expose the soil.
8. Hold a meterstick vertically with the 0 cm end touching the soil. Hold the three probes at the top of the meterstick. The light probe should be pointing straight down.
9. Select COLLECT DATA from the MAIN MENU. Select MONITOR INPUT from the DATA COLLECTION menu.
10. Wait for the readings of all three probes to stabilize. Then record their values in **Table 1.**
11. Move the probes down 10 cm to the 90 cm mark. Repeat step 10.
12. Continue recording the temperature, relative humidity, and light intensity at 10 cm intervals. For the last recording, hold the sensors as close to the 0 cm mark as you can without letting them touch the soil. Record all values in **Table 1.**

Name ______________________ Class ______________ Date ____________

Assessing Abiotic Factors in the Environment *continued*

TABLE 1 AIR TEMPERATURE, RELATIVE HUMIDITY, AND LIGHT INTENSITY

Height (cm)	Air temperature (°C)	Relative humidity (%)	Light intensity (μW/cm²)
100	**30.5**	**19.6**	**0.87**
90	**30.4**	**20.0**	**0.87**
80	**30.4**	**20.0**	**0.87**
70	**30.4**	**20.2**	**0.87**
60	**30.4**	**20.2**	**0.85**
50	**30.4**	**20.2**	**0.80**
40	**30.5**	**21.0**	**0.74**
30	**30.6**	**22.4**	**0.68**
20	**30.7**	**25.0**	**0.55**
10	**30.8**	**26.5**	**0.25**
0	**32.7**	**58.6**	**0.07**

Entries will vary for each group. Sample data are entered above.

DAY 1: MEASURING SOIL TEMPERATURE AND COLLECTING A SOIL SAMPLE

13. Push a large nail into the soil to a depth of 10 cm. Remove the nail.

14. Loosen the rubber band that holds the probes together, and insert the temperature probe all the way into the hole. When the temperature reading stabilizes, record the soil temperature in **Table 2.**

TABLE 2 SOIL TEMPERATURE, MOISTURE, AND pH

Soil temperature (°C)	**20.2**
Mass of soil before drying (g)	**287.6**
Mass of dried soil (g)	**227.7**
Mass of soil moisture (g)	**59.9**
Percentage of soil moisture	**20.8**
Soil pH	**6.3**

Entries will vary for each group. Sample data are entered above.

15. Turn off the CBL unit and the calculator.

16. Use a garden trowel to dig into the soil to a depth of 10 cm. Collect enough soil from that depth to fill a plastic sandwich bag about half full. Seal the bag, and return to the lab with the soil sample and the rest of your materials. Wash your hands thoroughly.

Name ______________________ Class ______________ Date ____________

DAY 1: MEASURING THE MASS OF THE SOIL SAMPLE

17. Place a sheet of filter paper over one end of a metal can with both ends removed, as shown in **Figure 1.** Secure the filter paper with a rubber band.

18. Use a balance to measure the mass of the soil you collected. Record the mass in **Table 2.**

19. Place the soil in the container you made in step 17. Use a permanent marker to label your container with the initials of each member of your group. Place your container in a drying oven set at 100°C for 24 hours.

FIGURE 1 SOIL CONTAINER

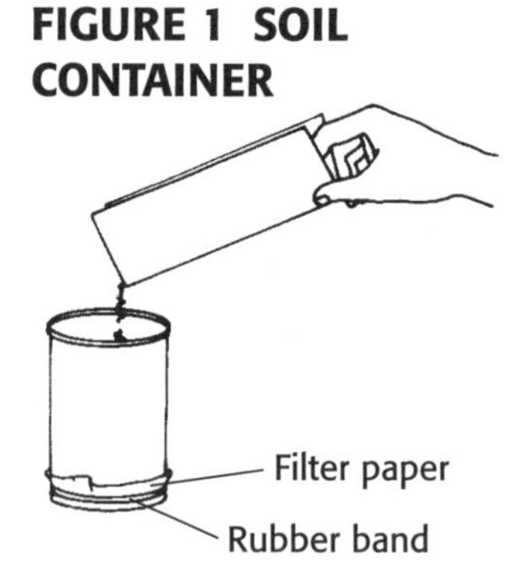

DAY 2: MEASURING THE MASS OF THE DRIED SOIL SAMPLE

20. Put on safety goggles and a lab apron.

21. Using oven mitts, carefully remove your container from the oven. **CAUTION: Do not touch the container with your bare hands.** Empty the dried soil from the container onto a paper plate. Use a metal spatula to dislodge any soil that sticks to the container.

22. Use a balance to measure the mass of the dried soil. Record the mass in **Table 2.**

DAY 2: PREPARING THE SOIL SAMPLE FOR pH MEASUREMENT

23. Use a sieve or strainer to remove large particles such as rocks from the dried soil. Grind 30 g of the sieved soil with a mortar and pestle.

24. Place the ground soil in a jar. **CAUTION: If a piece of glassware breaks, do not pick it up with your bare hands. Notify your teacher immediately.** Add 60 mL of deionized water to the jar. Attach the lid and shake the jar vigorously for 1 minute.

25. Filter the soil-water mixture, collecting the filtrate in a 100 mL beaker.

DAY 2: SETTING UP THE CBL SYSTEM

26. Connect the pH probe into the Channel 1 input of the CBL unit. Use the black cable to connect the CBL unit to the graphing calculator.

27. Turn on both the CBL unit and the calculator. Start the CHEMBIO program and go to the MAIN MENU.

28. Select SET UP PROBES. Enter “1” as the number of probes. Select pH from the SELECT PROBE menu. Enter “1” as the channel number.

29. Select USE STORED from the CALIBRATION menu.

30. Select COLLECT DATA from the MAIN MENU. Select MONITOR INPUT from the DATA COLLECTION menu.

Name ______________________________ Class ______________ Date ____________

Assessing Abiotic Factors in the Environment *continued*

DAY 2: MEASURING THE pH OF THE SOIL-WATER FILTRATE

31. Remove the pH probe from its storage solution. Carefully rinse the probe with deionized water, catching the rinse water in a 500 mL beaker.

32. Submerge the pH probe into the filtrate. When the pH reading stabilizes, record it in **Table 2.**

33. Turn off the CBL unit and the calculator. Rinse the pH probe thoroughly with deionized water, and return the probe to its storage solution, making sure that the cap is on tight. Dispose of the filtrate and the rinse water as instructed by your teacher.

34. Share the data you have recorded in **Table 1** and **Table 2** with the members of the other groups in your class.

35. Put away your materials, and clean up your work area. Wash your hands thoroughly before leaving the lab.

Analysis

1. Constructing Graphs On the graph in **Figure 2,** plot the air temperature versus the height above ground at the location you studied. Label the *y*-axis.

FIGURE 2 AIR TEMPERATURE VERSUS HEIGHT

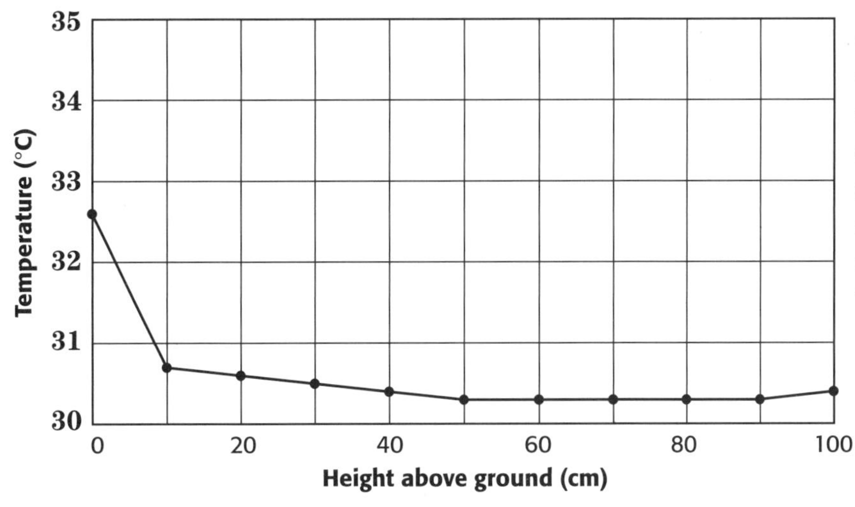

Graphs will vary, based on each group's data. The graph shown is based on the sample data given in Table 1.

Name ______________________ Class ______________ Date ____________

Assessing Abiotic Factors in the Environment *continued*

2. Constructing Graphs On the graph in **Figure 3,** plot the relative humidity versus the height above ground at the location you studied.

FIGURE 3 RELATIVE HUMIDITY VERSUS HEIGHT

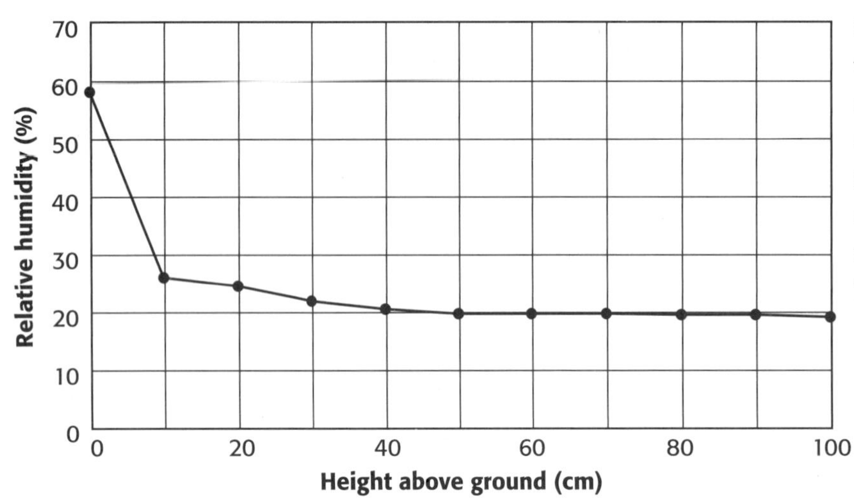

Graphs will vary, based on each group's data. The graph shown is based on the sample data given in Table 1.

3. Constructing Graphs On the graph in **Figure 4,** plot the light intensity versus the height above ground at the location you studied.

FIGURE 4 LIGHT INTENSITY VERSUS HEIGHT

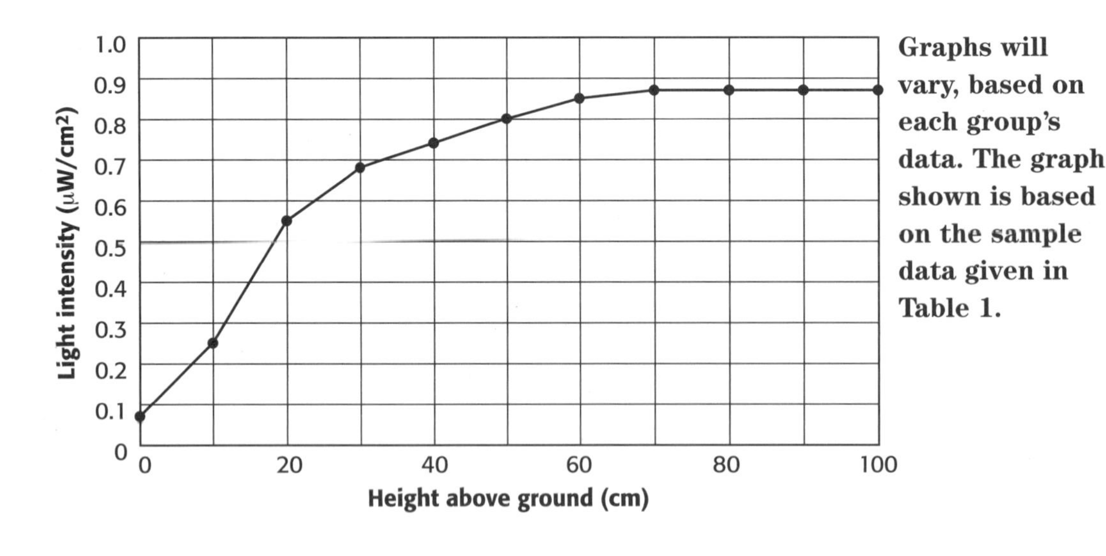

4. Organizing Data Calculate the mass of soil moisture in your sample by subtracting the mass of dried soil from the mass of soil before drying. Record the result in **Table 2.**

5. Organizing Data Calculate the percentage of soil moisture in your sample, using the following formula: percentage of soil moisture = (mass of soil moisture ÷ mass of soil before drying) × 100. Record the result in **Table 2.**

Name ______________________ Class ______________ Date ______________

Conclusions

1. **Analyzing Graphs** How did the air temperature, relative humidity, and light intensity vary with the height above ground at the location you studied?

   Answers will vary. In locations with direct sunlight and little or no ground vegetation, air temperature will likely remain fairly constant with height except very close to the ground, where it will likely be higher due to heat transfer from the soil. In shaded areas, the reverse may be true if the soil is cooler than the air. Unless the soil is very dry, the relative humidity is likely to be higher close to the ground. Light intensity should increase with height.

2. **Evaluating Methods** Why is it important to measure abiotic factors, such as air temperature, relative humidity, and light intensity, at different heights?

   If abiotic factors vary with height, that could affect the numbers and types of organisms that exist at different heights in the same location.

3. **Evaluating Results** Compare the air temperature, relative humidity, and light intensity at the location you studied with those abiotic factors at the location studied by another group.

   Answers will depend on the characteristics of the two locations that are compared. Locations with similar characteristics, such as ground cover and exposure to direct sunlight, should have similar abiotic factors. In shaded locations, the air temperature and light intensity should be lower and the relative humidity may be higher than in locations exposed to direct sunlight.

4. **Interpreting Information** How did the soil temperature at a depth of 10 cm compare with the air temperature at 0 cm? Account for any difference.

   Answers will vary. During warm, sunny weather, the air near the ground may absorb heat from the ground's surface, which is warmed by sunlight. The soil 10 cm below the surface does not receive sunlight, so it is likely to be cooler than the surface and the air just above the surface. During cold, windy weather under overcast skies, the surface of the ground may be cooled by the air. The soil below the surface is not in contact with the air, so it may be warmer than the surface and the air just above the surface.

Name ______________________ Class ______________ Date ______________

5. **Drawing Conclusions** Compare the percentage of soil moisture at the location you studied with that at the location studied by another group. Provide a reasonable explanation for any difference.

   Answers will depend on the two locations that are compared. There may be many reasons for a difference in soil moisture, including differences in soil composition, ground cover, shade, wind exposure, watering from sprinklers, and proximity to ponds or creeks. Accept any reasonable explanation.

6. **Making Predictions** If you were to plant a hydrangea at the location you studied, what would be the color of its bracts? Explain your reasoning.

   Students who found the soil to be acidic (pH less than 7) should predict that the bracts would be blue. Students who found the soil to be alkaline (pH greater than 7) should predict that the bracts would be pink.

Extensions

1. **Research and Communications** *Soil scientists* study all aspects of the soil, from its origin to its composition and distribution. A soil scientist may test the effects of fertilizer on soil quality, assess how a field drains, or develop practices to increase soil productivity. Conducting chemical analysis of soil and examining soil for bacterial and mineral content are other jobs that soil scientists perform. Find out what training and skills are needed to become a soil scientist. Report your findings in a short paper or oral presentation.
2. **Research and Communications** Select a plant and determine what it needs to thrive in any given region. List the abiotic factors that influence the survival of the plant. Present your list in a poster.
3. **Designing Experiments** Other important abiotic factors in terrestrial environments include soil nutrients and wind. Devise a procedure to measure these factors in different locations.

Answer Key

Directed Reading

SECTION: WHAT IS AN ECOSYSTEM?

1. f
2. a
3. e
4. b
5. d
6. c
7. c
8. b
9. c
10. a
11. d
12. d
13. b
14. Pioneer species are small, fast-growing plants that are the first organisms to live in a habitat.
15. Succession is a somewhat regular progression of species replacement. Succession on land where plants have not grown before is called primary succession. Succession on previously occupied land is called secondary succession.
16. The glacier in Glacier Bay has receded 100 km (62 mi) over the past 200 years. Thus, we can observe recently exposed land that contains few nutrients as well as land that has had more time to undergo succession.

SECTION: ENERGY FLOW IN ECOSYSTEMS

1. Producers are organisms that capture energy and use it to produce energy-storing molecules. Consumers obtain their energy by consuming producers or other consumers.
2. A trophic level identifies an organism's source of energy. A food chain is the path of energy through the trophic levels of an ecosystem.
3. Herbivores are animals that eat plants. Carnivores are animals that eat other animals.
4. Detritivores obtain their energy from organic wastes and dead organisms. Decomposers are fungi and bacteria that cause decay. Decomposers are one group of detritivores.
5. g
6. d
7. f
8. i
9. b
10. h
11. e
12. a
13. c
14. 10
15. energy pyramid
16. primary productivity

SECTION: CYCLING OF MATERIALS IN ECOSYSTEMS

1. cycles formed by the paths that materials take as they move between the living and nonliving parts of the environment
2. Living reservoirs are organisms within an ecosystem, whereas nonliving reservoirs are the physical components of an ecosystem, such as air and lakes.
3. water, carbon, nitrogen, and phosphorus
4. transpiration
5. ground water
6. the roots of plants
7. transpiration
8. Photosynthetic organisms use carbon dioxide in the air to form organic compounds.
9. cellular respiration, combustion, and decomposition
10. proteins, nucleic acids
11. calcium phosphate
12. nitrogen fixation
13. enzymes

Active Reading

SECTION: WHAT IS AN ECOSYSTEM?

1. biotic factors
2. biodiversity
3. ecology
4. community
5. ecosystem

6. abiotic factors
7. habitat
8. community
9. ecosystem
10. c

SECTION: ENERGY FLOW IN ECOSYSTEMS

1. A food chain is the path that energy takes as it moves through the trophic levels of an ecosystem.
2. producers
3. herbivores
4. carnivores and omnivores
5. Both omnivores and carnivores eat herbivores. However, omnivores also eat plants.
6. 2
7. 1
8. 3
9. 2
10. 3
11. 1
12. 3
13. 2
14. a

SECTION: CYCLING OF MATERIALS IN ECOSYSTEMS

1. Water vapor in the atmosphere condenses and falls to the Earth's surface as rain or snow.
2. Some becomes ground water, and the remaining precipitation reenters the atmosphere through evaporation.
3. Water is taken up by the roots of plants.
4. the process by which water moves into the atmosphere through tiny openings in the leaves of plants
5. In order for transpiration to occur, the sun must heat Earth's atmosphere to create the wind currents that draw moisture from plants.
6. water vapor
7. transpiration
8. evaporation
9. ground water
10. precipitation
11. d

Vocabulary Review

1. biodiversity
2. succession
3. ecosystem
4. biotic factors
5. pioneer species
6. primary succession
7. community
8. ecology
9. secondary succession
10. abiotic factors
11. habitat
12. f
13. d
14. i
15. k
16. m
17. q
18. p
19. o
20. l
21. n
22. a
23. g
24. c
25. b
26. e
27. j
28. h

Science Skills

INTERPRETING GRAPHICS

1–4. Answers may include any of the following food chains:

Chain 1—algae, krill, Adelie penguin, leopard seal, killer whale
Chain 2—algae, krill, cod, leopard seal, killer whale
Chain 3—algae, small animals and protists, krill, crabeater seal, killer whale
Chain 4—algae, small animals and protists, krill, Adelie penguin, killer whale
Chain 5—algae, small animals and protists, krill, crabeater seal, leopard seal, killer whale
Chain 6—algae, krill, crabeater seal, leopard seal, killer whale
Chain 7—algae, krill, crabeater seal, killer whale
Chain 8—algae, krill, adelie penguin, killer whale

5. krill, algae, small animals, and protists
6. leopard seal, killer whale, and elephant seal
7. one; algae
8. A
9. C
10. B
11. B; each higher level on the pyramid contains only 10 percent of the

biomass found in the trophic level below it. Ecologists measure biomass to determine the amount of energy present in a trophic level.

Concept Mapping

1. energy
2. biotic factors
3. abiotic factors
4. succession
5. water or nitrogen
6. carbon
7. soil
8. nitrogen or water
9. primary succession
10. secondary succession
11. food webs or food chains
12. food chains or food webs

Critical Thinking

1. d
2. c
3. a
4. b
5. b
6. d
7. a
8. c
9. c
10. a
11. d
12. b
13. f, a
14. c, i
15. e, d
16. h, b
17. g, j
18. a
19. b
20. d
21. c
22. d

Test Prep Pretest

1. d
2. a
3. d
4. c
5. b
6. a
7. c
8. b
9. d
10. b
11. c
12. e
13. a
14. d
15. abiotic factors
16. energy pyramid
17. biomass, number
18. nitrogen fixation
19. ammonification
20. abiotic factors such as soil, water, and weather
21. Energy is always lost when it is transferred from one trophic level to the next.
22. Water enters the atmosphere through evaporation and transpiration. It returns to Earth as precipitation.
23. assimilation, ammonification, nitrification, and denitrification
24. by breaking down organic molecules in cellular respiration
25. the remains of organisms that have been buried for thousands or millions of years and are transformed by heat and pressure into coal, oil, and natural gas

Quiz

SECTION: WHAT IS AN ECOSYSTEM?

1. b
2. c
3. d
4. a
5. d
6. d
7. a
8. e
9. b
10. c

SECTION: ENERGY FLOW IN ECOSYSTEMS

1. b
2. d
3. d
4. a
5. c
6. b
7. d
8. a
9. e
10. c

SECTION: CYCLING OF MATERIALS IN ECOSYSTEMS

1. d
2. c
3. b
4. a
5. c
6. c
7. e
8. d
9. b
10. a

Chapter Test (General)

1. b
2. c
3. a
4. b
5. d
6. b
7. b
8. c
9. d
10. d
11. a
12. b
13. d
14. c
15. a
16. d
17. e
18. a
19. c
20. b

Chapter Test (Advanced)

1. d
2. a
3. b
4. c
5. d
6. b
7. a
8. c
9. d
10. b
11. a
12. d
13. c
14. a
15. d
16. f
17. g
18. d
19. a
20. b
21. c
22. e
23. Without bacteria and fungi, dead organisms would not decompose, and the nutrients within their bodies would be unavailable to other living organisms.
24. As the glacier retreated, seeds and spores of pioneer species, including lichens and mosses, were carried in by wind and landed on exposed rock and gravel. After about 10 years, alder seeds were able to take root. Alder roots can fix nitrogen, so willows and cottonwood were able to grow in the richer soil.
25. Producers capture from the sun all the energy for an ecosystem.

Lesson Plan

Section: What Is an Ecosystem?

Pacing

Regular Schedule: **with lab(s):** N/A **without lab(s):** 3 days

Block Schedule: **with lab(s):** N/A **without lab(s):** 1 1/2 days

Objectives

1. **Distinguish** an ecosystem from a community.
2. **Describe** the diversity of a representative ecosystem.
3. **Sequence** the process of succession.

National Science Education Standards Covered

UNIFYING CONCEPTS AND PROCESSES

UCP1: Systems, order, and organization

UCP2: Evidence, models, and explanation

UCP4: Evolution and equilibrium

SCIENCE AS INQUIRY

SI1: Abilities necessary to do scientific inquiry

SI2: Understandings about scientific inquiry

SCIENCE IN PERSONAL AND SOCIAL PERSPECTIVES

SPSP2: Population growth

SPSP3: Natural resources

LIFE SCIENCE: INTERDEPENDENCE OF ORGANISMS

LSInter2: Energy flows through ecosystems in one direction, from photosynthetic organisms to herbivores to carnivores and decomposers.

LSInter3: Organisms both cooperate and compete in ecosystems.

LSInter4: Living organisms have the capacity to produce populations of infinite size, but environments and resources are finite.

LSInter5: Human beings live within the world's ecosystems.

LIFE SCIENCE: MATTER, ENERGY, AND ORGANIZATION IN LIVING SYSTEMS

LSMat2: The energy for life primarily comes from the sun.

LSMat5: The distribution and abundance of organisms and populations in ecosystems are limited by the availability of matter and energy and the ability of the ecosystem to recycle materials.

KEY
SE = Student Edition **TE** = Teacher Edition
CRF = Chapter Resource File

Block 1

CHAPTER OPENER *(45 minutes)*

_ **Quick Review,** SE. Students answer questions covered in previous sections of the textbook as preparation for the chapter content. **(GENERAL)**

_ **Reading Activity,** SE. Students decide if they agree or disagree with two statements related to the chapter topic. Then they read the chapter to see if their opinions are confirmed. **(GENERAL)**

_ **Using the Figure,** TE. Students answer questions about the chapter opener photograph. **(GENERAL)**

_ **Opening Activity**, Ecological Spheres, TE. Students discuss the inputs required in a sealed self-sustaining ecosystem. **(GENERAL)**

Block 2

FOCUS *(5 minutes)*

_ **Bellringer Transparency.** Use this transparency as students enter the classroom and find their seats. **(GENERAL)**

MOTIVATE *(10 minutes)*

_ **Activity**, Species Count, TE. Take students on a walk around your school or other area and have students keep count of all the different species of plants and animals you encounter. **(GENERAL)**

TEACH *(30 minutes)*

_ **Teaching Transparency, Section Outline.** Use this transparency to give students a framework for the information in this section. **(GENERAL)**

_ **Group Activity**, Biodiversity, TE. Students make an inventory of the organisms found in an area. Have students save this information to use to make a food web in the next section of the chapter. **(GENERAL)**

_ **Teaching Tip**, Detritivores, TE. Students make a list of organisms that would be considered ditritivores. **(GENERAL)**

HOMEWORK

_ **Active Reading Worksheet, What Is an Ecosystem?, CRF.** Students read a passage related to the section topic and answer questions. **(GENERAL)**

- **Directed Reading Worksheet, What Is an Ecosystem?, CRF.** Students complete the exercises in this worksheet to help them understand the material as they read the section. **(BASIC)**
- **Quick Lab,** Evaluating Biodiversity, SE. Students make simple observations and draw conclusions about biodiversity in an ecosystem. **(GENERAL)**
- **Datasheets for In-Text Labs, Evaluating Biodiversity, CRF.**

Block 3

TEACH *(30 minutes)*

- **Demonstration**, TE. Show the class a rock covered with lichens or mosses. Ask the students why these organisms are known as "pioneers." Point out the partnership between a fungus and an alga or a cyanobacterium in the lichens. **(BASIC)**
- **Teaching Tip**, Succession, TE. Students draw an illustrated timeline representing succession that occurs after a forest fire has burned all of the vegetation in an area. Starting at time 0, the land should be barren. **(GENERAL)**
- **Teaching Transparency, Ecological Succession at Glacier Bay.** Use this transparency to discuss succession. succession. Have students locate Glacier Bay, Alaska, on a map. **(GENERAL)**
- **Quick Lab,** Modeling Succession, SE. Students measure the pH of a jar of milk, observe the milk's appearance, and record the data in their tables over a period of 7 days. **(GENERAL)**
- **Datasheets for In-Text Labs, Modeling Succession, CRF.**

CLOSE *(10 minutes)*

- **Reteaching,** TE. Show students a picture of a forest ecosystem. Have them describe the events that would take place after a strong fire scorched the area. **(BASIC)**
- **Quiz,** TE. Students answer questions that review the section material. **(GENERAL)**

HOMEWORK

- **Section Review,** SE. Assign questions 1–5 for review, homework, or quiz. **(GENERAL)**
- **Quiz, CRF.** This quiz consists of ten multiple choice and matching questions that review the section's main concepts. **(BASIC) Also in Spanish.**

Other Resource Options

- **Internet Connect.** Students can research Internet sources about Ecosystems with SciLinks Code HX4066.
- **Internet Connect.** Students can research Internet sources about Biodiversity with SciLinks Code HX4020.

- **go.hrw.com.** For worksheets, videos, and other teaching aids related to this chapter, visit the HRW Web site and type in the keyword HX4 ECO.
- **Biology Interactive Tutor CD-ROM,** Unit 7 Ecosystem Dynamics. Students watch animations and other visuals as the tutor explains ecosystem dynamics. Students assess their learning with interactive activities.
- **CNN Science in the News, Video Segment 14 Tropical Reforestation.** This video segment is accompanied by a **Critical Thinking Worksheet**.
- **CNN Student News.** Find the latest news, lesson plans, and activities related to important scientific events at **cnnstudentnews.com**.

Lesson Plan

Section: Energy Flow in Ecosystems

Pacing

Regular Schedule:	**with lab(s):** N/A	**without lab(s):** 2 days
Block Schedule:	**with lab(s):** N/A	**without lab(s):** 1 day

Objectives

1. Distinguish between producers and consumers.
2. Compare food webs with food chains.
3. Describe why food chains are rarely longer than three or four links.

National Science Education Standards Covered

UNIFYING CONCEPTS AND PROCESSES

UCP1: Systems, order, and organization

UCP2: Evidence, models, and explanation

SCIENCE AS INQUIRY

SI1: Abilities necessary to do scientific inquiry

SI2: Understandings about scientific inquiry

SCIENCE IN PERSONAL AND SOCIAL PERSPECTIVES

SPSP3: Natural resources

LIFE SCIENCE: INTERDEPENDENCE OF ORGANISMS

LSInter1: The atoms and molecules on Earth cycle among the living and nonliving components of the biosphere.

LSInter2: Energy flows through ecosystems in one direction, from photosynthetic organisms to herbivores to carnivores and decomposers.

LSInter3: Organisms both cooperate and compete in ecosystems.

LIFE SCIENCE: MATTER, ENERGY, AND ORGANIZATION IN LIVING SYSTEMS

LSMat1: All matter tends toward more disorganized states.

LSMat2: The energy for life primarily comes from the sun.

LSMat3: The chemical bonds of food molecules contain energy.

LSMat4: The complexity and organization of organisms accommodates the need for obtaining, transforming, transporting, releasing, and eliminating the matter and energy used to sustain the organism.

PHYSICAL SCIENCE

PS5: Conservation of energy and increase in disorder

PS6: Interactions of energy and matter

EARTH SCIENCE

ES1: Energy in the Earth system

KEY
SE = Student Edition
TE = Teacher Edition
CRF = Chapter Resource File

Block 4

FOCUS *(5 minutes)*

_ **Bellringer Transparency.** Use this transparency as students enter the classroom and find their seats. **(GENERAL)**

MOTIVATE *(10 minutes)*

_ **Demonstration**, TE. List the following organisms that can be found in an open field: robin, hawk, snake, frog, grasshopper, mouse, and rabbit. Have students draw arrows to show what eats what in this field ecosystem. **(GENERAL)**

TEACH *(30 minutes)*

_ **Teaching Transparency, Section Outline.** Use this transparency to give students a framework for the information in this section. **(GENERAL)**

_ **Teaching Transparency, Trophic Levels.** Use this transparency to discuss how energey moves from one trophic level to another. **(GENERAL)**

_ **Teaching Transparency, Food Chain in an Antarctic Ecosystem.** Use this transparency to display an example of a food chain. Trace the paths of energy through this food chain. Have students identify the organisms in each trophic level, beginning with the producers. **(GENERAL)**

_ **Teaching Tip,** Trophic Levels, TE. Students identify the trophic level of another student as he or she ate each item for dinner the night before. **(GENERAL)**

_ **Teaching Transparency, Food Web in an Antarctic Ecosystem.** Use this transparency to display an example of a food web. Emphasize the interconnectedness of this ecosystem. **(GENERAL)**

HOMEWORK

_ **Directed Reading Worksheet, Energy Flow in Ecosystems, CRF.** Students complete the exercises in this worksheet to help them understand the material as they read the section. **(BASIC)**

_ **Active Reading Worksheet, Energy Flow in Ecosystems, CRF.** Students read a passage related to the section topic and answer questions. **(GENERAL)**

Block 5

TEACH *(35 minutes)*

_ **Quick Lab, Making a Food Web, CRF.** Students draw food chains that might be found in a woodland community and show how the food chains are connected to form a food web. **(GENERAL)**

_ **Teaching Transparency, Energy Transfer Through Trophic Levels.** Use this transparency to point out that each trophic level contains about 90 percent less energy than the level below it. **(GENERAL)**

_ **Teaching Transparency, Energy Efficiency in Food Consumption.** Use this transparency to discuss human food choices. Point out that adding a trophic level increaes the energy demand by consumers by a factor of 10. **(GENERAL)**

_ **Activity,** Energy in Trophic Levels, TE. Tell students that if an average of 1,500 kilocalories of light energy per day falls on a square meter of land surface covered by plants, only about 15–30 kilocalories become incorporated into chemical compounds by photosynthesis. Ask students how much of this energy could end up in a person who eats plants versus a steak from a steer that ate plants. **(GENERAL)**

CLOSE *(10 minutes)*

_ **Reteaching,** TE. Students construct a food web and energy pyramid for the school yard and analyze the types of organisms present. **(BASIC)**

HOMEWORK

_ **Alternative Assessment**, TE. Students draw a food chain, a food web, and an energy pyramid for any ecosystem. Tell them to represent several trophic levels and feeding pathways. **(GENERAL)**

_ **Quiz,** TE. Students answer questions that review the section material. **(GENERAL)**

_ **Quiz, CRF.** This quiz consists of ten multiple choice and matching questions that review the section's main concepts. **(BASIC) Also in Spanish.**

_ **Section Review,** SE. Assign questions 1–6 for review, homework, or quiz. **(GENERAL)**

Other Resource Options

_ **Teaching Tip,** Physics, TE. Remind students that the first law of thermodynamics states that matter cannot be created or destroyed, but only changed from one form into another form. The second law states that energy conversion between different forms is never 100 percent efficient. Have students think of examples of each law. **(GENERAL)**

_ **Internet Connect.** Students can research Internet sources about Food Chains with SciLinks Code HX4085.

_ **Internet Connect.** Students can research Internet sources about Energy Pyramids with SciLinks Code HX4069.

_ **go.hrw.com.** For worksheets, videos, and other teaching aids related to this chapter, visit the HRW Web site and type in the keyword HX4 ECO.

_ **Biology Interactive Tutor CD-ROM,** Unit 7 Ecosystem Dynamics**.** Students watch animations and other visuals as the tutor explains ecosystem dynamics. Students assess their learning with interactive activities.

_ **CNN Science in the News, Video Segment 13 Fat Wolves.** This video segment is accompanied by a **Critical Thinking Worksheet**.

_ **CNN Student News.** Find the latest news, lesson plans, and activities related to important scientific events at **cnnstudentnews.com**.

Lesson Plan

Section: Cycling of Materials in Ecosystems

Pacing

Regular Schedule:	**with lab(s):** 4 days	**without lab(s):** 2 days
Block Schedule:	**with lab(s):** 2 days	**without lab(s):** 1 day

Objectives

1. Summarize the role of plants in the water cycle.
2. Analyze the flow of energy through the carbon cycle.
3. Identify the role of bacteria in the nitrogen cycle.

National Science Education Standards Covered

UNIFYING CONCEPTS AND PROCESSES

UCP1: Systems, order, and organization

UCP4: Evolution and equilibrium

SCIENCE AS INQUIRY

SI1: Abilities necessary to do scientific inquiry

SI2: Understandings about scientific inquiry

LIFE SCIENCE: INTERDEPENDENCE OF ORGANISMS

LSInter1: The atoms and molecules on Earth cycle among the living and nonliving components of the biosphere.

LSInter2: Energy flows through ecosystems in one direction, from photosynthetic organisms to herbivores to carnivores and decomposers.

LIFE SCIENCE: MATTER, ENERGY, AND ORGANIZATION IN LIVING SYSTEMS

LSMat1: All matter tends toward more disorganized states.

LSMat2: The energy for life primarily comes from the sun.

LSMat3: The chemical bonds of food molecules contain energy.

LSMat4: The complexity and organization of organisms accommodates the need for obtaining, transforming, transporting, releasing, and eliminating the matter and energy used to sustain the organism.

PHYSICAL SCIENCE

PS5: Conservation of energy and increase in disorder

PS6: Interactions of energy and matter

EARTH SCIENCE

ES1: Energy in the Earth system

ES2: Geochemical cycles

KEY

SE = Student Edition **TE** = Teacher Edition
CRF = Chapter Resource File

Block 6

FOCUS *(5 minutes)*

_ **Bellringer Transparency.** Use this transparency as students enter the classroom and find their seats. **(GENERAL)**

MOTIVATE *(10 minutes)*

_ **Discussion**, TE. Use the discussion scenario and question from this TE item to generate a discussion about the water cycle. **(GENERAL)**

TEACH *(30 minutes)*

_ **Teaching Transparency, Section Outline.** Use this transparency to give students a framework for the information in this section. **(GENERAL)**

_ **Teaching Transparency, Water Cycle.** Use this transparency to discuss the water cycle. Review the terms *precipitation*, *runoff*, *transpiration*, and *evaporation*. Trace the path that a molecule of water follows through the cycle. **(GENERAL)**

_ **Teaching Tip**, Atmospheric Carbon Dioxide, TE. Students make a list of activities that contribute to rising levels of carbon dioxide. **(GENERAL)**

_ **Teaching Transparency, Carbon Cycle.** Use this transparency to describe the the carbon cycle. Discuss the impact of burning fossil fuels on the climate. Have students list three ways humans add carbon to the atmosphere. **(GENERAL)**

HOMEWORK

_ **Directed Reading Worksheet, Cycling of Materials in Ecosystems, CRF.** Students complete the exercises in this worksheet to help them understand the material as they read the section. **(BASIC)**

_ **Active Reading Worksheet, Cycling of Materials in Ecosystems, CRF.** Students read a passage related to the section topic and answer questions. **(GENERAL)**

Block 7

TEACH *(30 minutes)*

_ **Demonstration**, TE. Show students the nodules on a leguminous plant. **(BASIC)**

_ **Teaching Transparency, Nitrogen Cycle.** Use this transparency to describe the nitrogen cycle. Review the terms *nitrogen fixation, ammonification, nitrification,* and *denitrification.* Have students explain why nitrogen is scarce in the soil even though the atmosphere is 78 percent nitrogen gas. **(GENERAL)**

_ **Using the Figure**, Figure 14, TE. Help students distinguish between nitrogen fixation and nitrification. Tell students that lightning also changes nitrogen gas into ammonia, but it amounts to less than 10 percent of that carried out by organisms through nitrogen fixation. **(GENERAL)**

_ **Teaching Tip**, Gene Splicing, TE. Students discuss the pros and cons of putting a nitrogen-fixing gene in non-leguminous plants. **(GENERAL)**

CLOSE *(15 minutes)*

_ **Reteaching,** TE. Students write a short essay about what the world would be like without bacteria or fungi. **(BASIC)**

_ **Alternative Assessment,** TE. Students write an essay about how humans would be affected if carbon or nitrogen did not cycle. **(ADVANCED)**

HOMEWORK

_ **Quiz,** TE. Students answer questions that review the section material. **(GENERAL)**

_ **Section Review,** SE. Assign questions 1–5 for review, homework, or quiz. **(GENERAL)**

_ **Science Skills Worksheet, CRF.** Students interpret a graphic of a food web and graphics of energy pyramids. **(GENERAL)**

_ **Quiz, CRF.** This quiz consists of ten multiple choice and matching questions that review the section's main concepts. **(BASIC) Also in Spanish.**

_ **Modified Worksheet, One-Stop Planner.** This worksheet has been specially modified to reach struggling students. **(BASIC)**

_ **Critical Thinking Worksheet, CRF.** Students answer analogy-based questions that review the section's main concepts and vocabulary. **(ADVANCED)**

Optional Blocks

LAB *(90 minutes)*

_ **Skills Practice Lab, Assessing Abiotic Factors in the Environment, CRF.** Students measure six abiotic factors in the environment at one location on the grounds of the school. They then compare their findings with those of groups that have measured abiotic factors at other locations. **(GENERAL)**

Other Resource Options

- **Exploration Lab,** Modeling Ecosystem Change over Time, SE. Students model a natural ecosystem by building an ecosystem in a jar. They then observe the interactions of the organisms in the ecosystem and note any changes that occur over a period of a few weeks. **(GENERAL)**
- **Datasheets for In-Text Labs, Modeling Ecosystem Change over Time, CRF.**
- **Internet Connect.** Students can research Internet sources about Water Cycle with SciLinks Code HX4188.
- **go.hrw.com.** For worksheets, videos, and other teaching aids related to this chapter, visit the HRW Web site and type in the keyword HX4 ECO.
- **Biology Interactive Tutor CD-ROM,** Unit 7 Ecosystem Dynamics. Students watch animations and other visuals as the tutor explains ecosystem dynamics. Students assess their learning with interactive activities.
- **CNN Science in the News, Video Segment 14 Tropical Reforestation.** This video segment is accompanied by a **Critical Thinking Worksheet**.
- **CNN Student News.** Find the latest news, lesson plans, and activities related to important scientific events at **cnnstudentnews.com**.

Lesson Plan

End-of-Chapter Review and Assessment

Pacing

Regular Schedule: 2 days

Block Schedule: 1 day

KEY	
SE = Student Edition	**TE** = Teacher Edition
CRF = Chapter Resource File	

Block 8

REVIEW *(45 minutes)*

- **Study Zone,** SE. Use the Study Zone to review the Key Concepts and Key Terms of the chapter and prepare students for the Performance Zone questions. **(GENERAL)**
- **Performance Zone,** SE. Assign questions to review the material for this chapter. Use the assignment guide to customize review for sections covered. **(GENERAL)**
- **Teaching Transparency, Concept Mapping.** Use this transparency to review the concept map for this chapter. **(GENERAL)**

Block 9

ASSESSMENT *(45 minutes)*

- **Chapter Test, Ecosystems, CRF.** This test contains 20 multiple choice and matching questions keyed to the chapter's objectives. **(GENERAL) Also in Spanish.**
- **Chapter Test, Ecosystems, CRF.** This test contains 25 questions of various formats, each keyed to the chapter's objectives. **(ADVANCED)**
- **Modified Chapter Test, One-Stop Planner.** This test has been specially modified to reach struggling students. **(BASIC)**

Other Resource Options

- **Vocabulary Review Worksheet, CRF.** Use this worksheet to review the chapter vocabulary. **(GENERAL) Also in Spanish.**
- **Test Prep Pretest, CRF.** Use this pretest to review the main content of the chapter. Each question is keyed to a section objective. **(GENERAL) Also in Spanish.**
- **Test Item Listing for ExamView® Test Generator, CRF.** Use the Test Item Listing to identify questions to use in a customized homework, quiz, or test.
- **ExamView® Test Generator, One-Stop Planner.** Create a customized homework, quiz, or test using the HRW Test Generator program.

Ecosystems

TRUE/FALSE

1. ____ Ecologists call the physical location of a community its habitat.
Answer: True Difficulty: I Section: 1 Objective: 1

2. ____ Ecosystems include only the biotic factors in an area.
Answer: False Difficulty: I Section: 1 Objective: 1

3. ____ A community includes all the species within an area.
Answer: True Difficulty: I Section: 1 Objective: 1

4. ____ Biotic factors in a habitat include all the physical aspects as well as the living organisms.
Answer: False Difficulty: I Section: 1 Objective: 1

5. ____ Biotic factors of a habitat include all abiotic factors.
Answer: False Difficulty: I Section: 1 Objective: 1

6. ____ The number of species living within an ecosystem is a measure of its biodiversity.
Answer: True Difficulty: I Section: 1 Objective: 2

7. ____ The major difference between primary succession and secondary succession is that primary succession occurs only on land and secondary succession occurs in ponds and lakes.
Answer: False Difficulty: I Section: 1 Objective: 3

8. ____ Cutting down trees in a forest alters the habitat of the organisms living in the forest.
Answer: True Difficulty: I Section: 1 Objective: 3

9. ____ When succession takes place in an area where there has been previous growth, it is called secondary succession.
Answer: True Difficulty: I Section: 1 Objective: 3

10. ____ The first organisms to live in a new habitat are large, slow-growing plants.
Answer: False Difficulty: I Section: 1 Objective: 3

11. In a new habitat, pioneer species are eventually replaced by other plant immigrants.
Answer: True Difficulty: I Section: 1 Objective: 3

12. ____ A receding glacier is a good example of secondary succession.
Answer: False Difficulty: I Section: 1 Objective: 3

13. ____ Glaciers, like the ones in Alaska, move very slowly and are good for showing changes that have taken place as time passes.
Answer: True Difficulty: I Section: 1 Objective: 3

14. ____ When an organism dies, the nutrients in its body are released back into the environment by decomposers.
Answer: True Difficulty: I Section: 2 Objective: 1

15. ____ Decomposers absorb energy from organisms by breaking down living tissue.
Answer: False Difficulty: I Section: 2 Objective: 1

16. ____ The lowest trophic level of any ecosystem is occupied by the producers.
Answer: True Difficulty: I Section: 2 Objective: 1

17. ____ Producers in an ecosystem transfer all of their energy to primary-level consumers.
Answer: False Difficulty: I Section: 2 Objective: 1

18. ____ The source of energy for an organism determines its trophic level.
Answer: True Difficulty: I Section: 2 Objective: 1

19. ____ A trophic level is made up of a group of organisms whose energy sources are the same energy level away from the sun.
Answer: True Difficulty: I Section: 2 Objective: 1

20. ____ Omnivores feed only on primary producers.
Answer: False Difficulty: I Section: 2 Objective: 1

21. ____ Detritivores are especially harmful to an ecosystem.
Answer: False Difficulty: I Section: 2 Objective: 1

22. ____ A food chain is made up of interrelated food webs.
Answer: False Difficulty: I Section: 2 Objective: 2

23. ____ Food chains usually begin with the primary producers.
Answer: True Difficulty: I Section: 2 Objective: 2

24. ____ All organisms in an ecosystem are part of the food web of that ecosystem.
Answer: True Difficulty: I Section: 2 Objective: 2

25. ____ A change in the number of predators in a food web can affect an entire ecosystem.
Answer: True Difficulty: I Section: 2 Objective: 3

26. ____ The number of organisms in a trophic level is always directly proportional to the amount of energy at that level.
Answer: False Difficulty: I Section: 2 Objective: 3

27. ____ Organisms at higher trophic levels tend to be fewer in number than those at lower trophic levels.
Answer: True Difficulty: I Section: 2 Objective: 3

28. ____ Plants release water into the atmosphere through transpiration.
Answer: True Difficulty: I Section: 3 Objective: 1

29. ____ Water and nutrients continue to cycle normally in a forest ecosystem after the trees in the ecosystem have been cut down.
Answer: False Difficulty: I Section: 3 Objective: 1

30. ____ Transpiration is a sun-driven process.
Answer: True Difficulty: I Section: 3 Objective: 1

31. ____ Carbon is returned to the atmosphere by photosynthesis, combustion, and erosion.
Answer: False Difficulty: I Section: 3 Objective: 2

32. ____ Nitrogen gas makes up about 79 percent of the Earth's atmosphere.
Answer: True Difficulty: I Section: 3 Objective: 3

33. ____ During nitrification, decomposers break down the roots of plants to produce nitrates.
Answer: False Difficulty: I Section: 3 Objective: 3

MULTIPLE CHOICE

34. A functioning aquarium displays
a. a community.
b. a habitat.
c. an ecosystem.
d. All of the above
Answer: D Difficulty: I Section: 1 Objective: 1

35. A group of organisms of different species living together in a particular place is called a
    a. community.
    b. population.
    c. biome.
    d. habitat.

    Answer: A   Difficulty: I   Section: 1   Objective: 1

36. The physical location of an ecosystem in which a given species lives is called a
    a. habitat.
    b. tropical level.
    c. community.
    d. food zone.

    Answer: A   Difficulty: I   Section: 1   Objective: 1

37. Ecology is the study of the interaction of living organisms
    a. with one another and their biotic factors.
    b. and their communities.
    c. with one another and their physical environment.
    d. and the food they eat.

    Answer: C   Difficulty: I   Section: 1   Objective: 1

38. An ecosystem consists of
    a. a community of organisms.
    b. energy.
    c. the soil, water, and weather.
    d. All of the above

    Answer: D   Difficulty: I   Section: 1   Objective: 1

39. All of the following are abiotic factors of a habitat *except*
    a. soil.
    b. plants.
    c. water.
    d. weather.

    Answer: B   Difficulty: I   Section: 1   Objective: 1

40. Biodiversity measures the number of species living within a(n)
    a. ecosystem.
    b. habitat.
    c. organism.
    d. community.

    Answer: A   Difficulty: I   Section: 1   Objective: 2

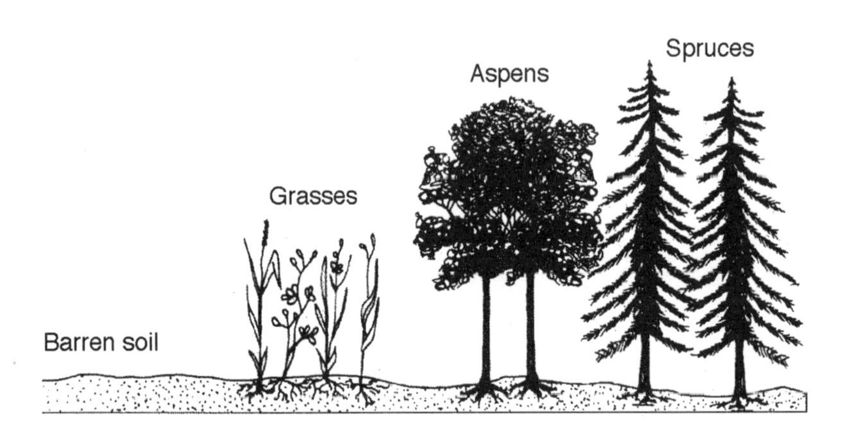

41. Refer to the illustration above. The process shown in the diagram is known as
    a. competitive exclusion.
    b. succession.
    c. symbiosis.
    d. oligotrophy.

    Answer: B   Difficulty: II   Section: 1   Objective: 3

42. Succession is
    a. an organism's ability to survive in its environment.
    b. the number of species living in an ecosystem.
    c. the regular progression of species replacement in an environment.
    d. the transfer of energy through a food chain.

    Answer: C Difficulty: I Section: 1 Objective: 3

43. Which of the following types of succession would most likely occur following a forest fire?
    a. primary succession
    b. old field succession
    c. secondary succession
    d. lake succession

    Answer: C Difficulty: I Section: 1 Objective: 3

44. Secondary succession occurs
    a. as one generation of organisms replaces the previous one.
    b. as a previously existing community is replaced.
    c. after a new food web is established.
    d. None of the above

    Answer: B Difficulty: I Section: 1 Objective: 3

45. When the settlers arrived in New England, many forests were turned into fields. Eventually, some fields were abandoned and then grew back into forests. This is best described as
    a. primary succession.
    b. coevolution.
    c. secondary succession.
    d. niche realization.

    Answer: C Difficulty: I Section: 1 Objective: 3

46. primary succession : areas of no previous plant growth ::
    a. new habitat : a climax community
    b. rain forest : a desert
    c. tundra : a desert
    d. secondary succession : abandoned farm fields

    Answer: D Difficulty: II Section: 1 Objective: 3

47. When an organism dies, the nutrients in its body
    a. can never be reused by other living things.
    b. are immediately released into the atmosphere.
    c. are released by the action of decomposers.
    d. None of the above

    Answer: C Difficulty: I Section: 2 Objective: 1

48. Fungi are
    a. decomposers.
    b. scavengers.
    c. omnivores.
    d. autotrophs.

    Answer: A Difficulty: I Section: 2 Objective: 1

49. A relationship between a producer and consumer is best illustrated by a
    a. snake eating a bird.
    b. fox eating a mouse.
    c. lion eating a zebra.
    d. zebra eating grass.

    Answer: D Difficulty: I Section: 2 Objective: 1

50. Organisms that manufacture organic nutrients for an ecosystem are called
    a. primary consumers.
    b. predators.
    c. primary producers.
    d. scavengers.

    Answer: C Difficulty: I Section: 2 Objective: 1

51. The primary producers in a grassland ecosystem would most likely be
    a. insects.
    b. bacteria.
    c. grasses.
    d. algae.

    Answer: C Difficulty: I Section: 2 Objective: 1

52. cows : herbivores ::
    a. horses : carnivores
    b. plants : producers
    c. algae : consumers
    d. caterpillars : producers

    Answer: B Difficulty: II Section: 2 Objective: 1

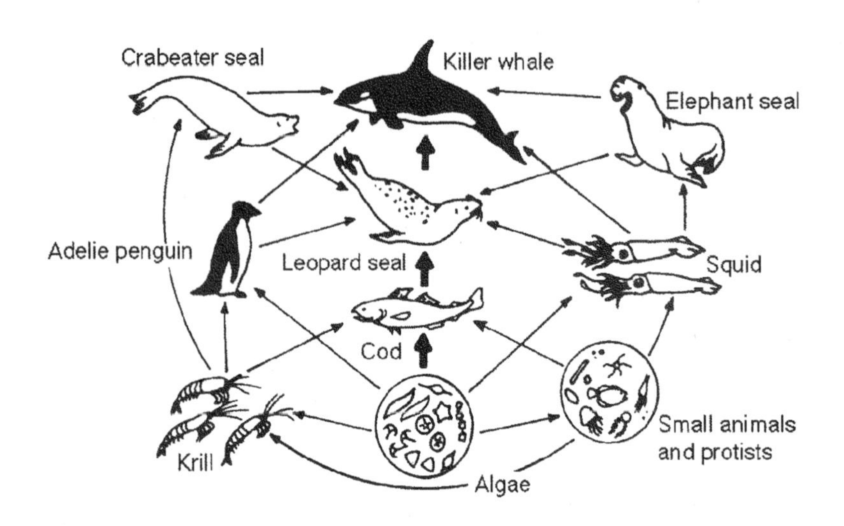

53. Refer to the illustration above. The photosynthetic algae are
    a. producers.
    b. consumers.
    c. parasites.
    d. decomposers.

    Answer: A Difficulty: II Section: 2 Objective: 1

54. Refer to the illustration above. The diagram, which shows how energy moves through an ecosystem, is known as a
    a. habitat.
    b. food chain.
    c. food net.
    d. food web.

    Answer: D Difficulty: II Section: 2 Objective: 2

55. Refer to the illustration above. Leopard seals are
    a. producers.
    b. omnivores.
    c. herbivores.
    d. carnivores.

    Answer: D Difficulty: II Section: 2 Objective: 2

56. Refer to the illustration above. Killer whales feed at the
    a. first and second trophic levels.
    b. second trophic level only.
    c. second and third trophic levels.
    d. third and fourth trophic levels.

    Answer: D Difficulty: II Section: 2 Objective: 2

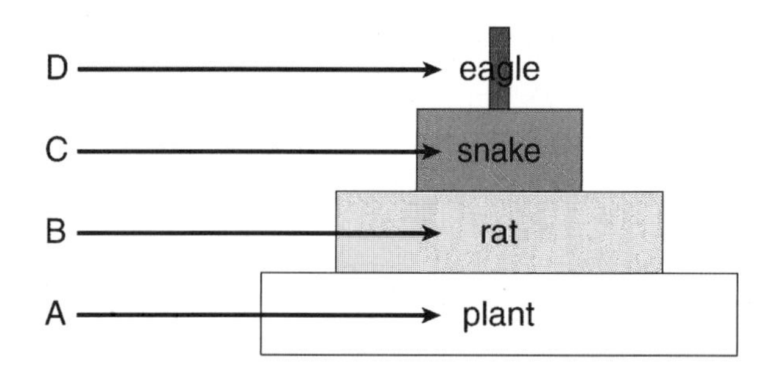

57. Refer to the illustration above. Level A is composed of
    a. carnivores.
    b. herbivores.
    c. producers.
    d. omnivores.

    Answer: C Difficulty: II Section: 2 Objective: 2

58. Refer to the illustration above. The diagram shows a(n)
    a. food chain.
    b. community.
    c. food web.
    d. energy pyramid.

    Answer: D Difficulty: II Section: 2 Objective: 3

59. Refer to the illustration above. On the pyramid, animals that feed on plant eaters are no lower than
    a. level A.
    b. level B.
    c. level C.
    d. level D.

    Answer: C Difficulty: II Section: 2 Objective: 3

60. Refer to the illustration above. How much energy is available to the organisms in level C?
    a. all of the energy in level A plus the energy in level B
    b. all of the energy in level A minus the energy in level B
    c. 10 percent of the energy in level B
    d. 90 percent of the energy in level B

    Answer: C Difficulty: II Section: 2 Objective: 3

61. Refer to the illustration above. The diagram represents the decrease in
    a. the number of organisms between lower and higher trophic levels.
    b. available energy between lower and higher trophic levels.
    c. diversity of organisms between lower and higher levels.
    d. All of the above

    Answer: D Difficulty: II Section: 2 Objective: 3

62. Food webs are more commonplace than food chains because
    a. many animals that comprise the links in a food chain are migratory.
    b. organisms almost always eat, and are eaten by, many different organisms.
    c. over time, food chains always become food webs.
    d. None of the above

    Answer: B Difficulty: I Section: 2 Objective: 2

63. In a food web, which type of organism receives energy from every other type?
    a. producer
    b. carnivore
    c. decomposer
    d. herbivore

    Answer: C Difficulty: I Section: 2 Objective: 2

64. Animals that feed on plants are at least in the
    a. first trophic level.
    b. second trophic level.
    c. third trophic level.
    d. fourth trophic level.

    Answer: B Difficulty: I Section: 2 Objective: 2

65. Which of the following are detritivores?
    a. worms
    b. vultures
    c. fungi
    d. All of the above

    Answer: D Difficulty: I Section: 2 Objective: 2

66. The number of trophic levels in an ecological pyramid
    a. is limitless.
    b. is limited by the amount of energy that is lost at each trophic level.
    c. never exceeds four.
    d. never exceeds three.

    Answer: B Difficulty: I Section: 2 Objective: 3

67. In going from one trophic level to the next higher level,
    a. the number of organisms increases.
    b. the amount of usable energy increases.
    c. the amount of usable energy decreases.
    d. diversity of organisms increases.

    Answer: C Difficulty: I Section: 2 Objective: 3

68. The total dry weight of the organisms in an ecosystem is called
    a. trophic level.
    b. biomass.
    c. energy level.
    d. ecomass.

    Answer: B Difficulty: I Section: 2 Objective: 3

69. Because energy diminishes at each successive trophic level, few ecosystems can contain more than
    a. two trophic levels.
    b. four trophic levels.
    c. five trophic levels.
    d. eight trophic levels.

    Answer: B Difficulty: I Section: 2 Objective: 3

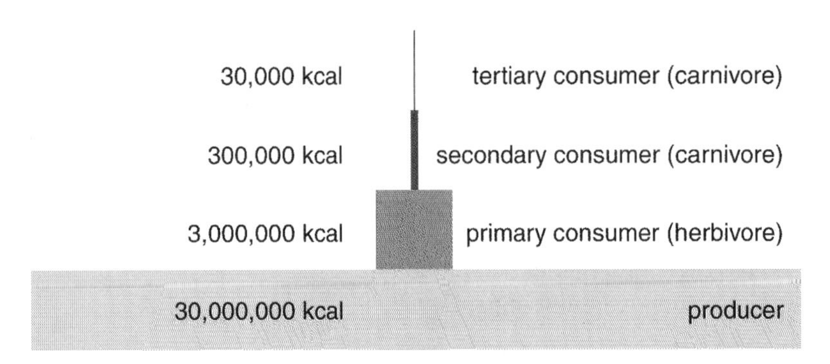

70. Refer to the illustration above. At each trophic level, the energy stored in the organisms in that level is
    a. about one-tenth of the energy in the level below it.
    b. about one-tenth of the energy in the level above it.
    c. 50 percent of the energy in the level below it.
    d. 100 percent of the energy in the level below it.

    Answer: A Difficulty: II Section: 2 Objective: 3

71. Precipitation and evaporation are important components of the
    a. nitrogen cycle.
    b. water cycle.
    c. carbon cycle.
    d. All of the above

    Answer: B Difficulty: I Section: 3 Objective: 1

72. The paths of water, carbon, nitrogen, and phosphorus pass from the nonliving environment to living organisms and back to the nonliving environment in closed circles called
    a. living cycles.
    b. environcycles.
    c. biogeochemical cycles.
    d. None of the above

    Answer: C Difficulty: I Section: 3 Objective: 1

73. Coal, oil, and natural gas
    a. are formed from decayed plants.
    b. are fossil fuels.
    c. release carbon dioxide when they are burned.
    d. All of the above

    Answer: D Difficulty: I Section: 3 Objective: 2

74. Humans are affecting the carbon cycle by
    a. burning fossil fuels.
    b. destroying vegetation that absorbs carbon dioxide.
    c. using electrical labor-saving devices.
    d. All of the above

    Answer: D Difficulty: I Section: 3 Objective: 2

75. Which of the following is part of the nitrogen cycle?
    a. conversion of atmospheric nitrogen into usable organic compounds by bacteria
    b. conversion of nitrogen from decaying organisms into ammonia
    c. nitrogen fixation
    d. All of the above

    Answer: D Difficulty: I Section: 3 Objective: 3

76. Nitrogen is a component of
    a. proteins.
    b. fats.
    c. carbohydrates.
    d. water.

    Answer: A Difficulty: I Section: 3 Objective: 3

77. denitrification : nitrogen gas in the atmosphere ::
    a. more rain : transformation of rain forests
    b. more transpiration : arid weather
    c. burning fossil fuels : carbon in the atmosphere
    d. combustion : ground water

    Answer: C Difficulty: II Section: 3 Objective: 3

78. ammonification : ammonia ::
    a. denitrification : nitrogen gas
    b. oil : gasoline
    c. nitrification : ammonia
    d. nitrification : oxygen

    Answer: A Difficulty: II Section: 3 Objective: 3

79. Which of the following is common to the carbon cycle, the nitrogen cycle, and the water cycle?
    a. The substance is rearranged into different types of molecules as it moves through its cycle.
    b. The substance must pass through organisms in order to complete its cycle.
    c. The largest reserves of the substance are always in organisms.
    d. The substance is required by all living things and is involved in many processes that occur in all living things.

    Answer: D Difficulty: II Section: 3 Objective: 3

COMPLETION

80. An ecosystem consists of the living and ____________________ environment.
Answer: nonliving Difficulty: I Section: 1 Objective: 1

81. The physical area in which an organism lives is its ____________________.
Answer: habitat Difficulty: I Section: 1 Objective: 1

82. Ernst Haeckel coined the term ____________________ to describe the study of how organisms interact with one another and with their environment.
Answer: ecology Difficulty: I Section: 1 Objective: 1

83. The living organisms in a habitat are called ____________________ factors.
Answer: biotic Difficulty: I Section: 1 Objective: 1

84. The number of species living within an ecosystem is a measure of its ____________________.
Answer: biodiversity Difficulty: I Section: 1 Objective: 2

85. The sequential replacement of populations in an area that has not previously supported life is called ____________________ succession.
Answer: primary Difficulty: I Section: 1 Objective: 3

86. The small, fast-growing plants that are the first organisms to live in a habitat are called ____________________ ____________________.
Answer: pioneer species Difficulty: I Section: 1 Objective: 3

87. A receding glacier is a good example of ____________________ ____________________.
Answer: primary succession Difficulty: I Section: 1 Objective: 3

88. Everything that organisms do in ecosystems, such as running, breathing, burrowing, and growing, requires ____________________.
Answer: energy Difficulty: I Section: 2 Objective: 1

89. Animals known as ____________________ eat only primary producers.
Answer: herbivores Difficulty: I Section: 2 Objective: 1

90. An organism that eats a primary consumer is called a(n) ____________________ consumer.
Answer: secondary Difficulty: I Section: 2 Objective: 1

91. The term ____________________ is given to the bacteria that break down dead tissue.
Answer: decomposers Difficulty: I Section: 2 Objective: 1

92. The primary productivity of an ecosystem is a measure of the amount of organic material that the ____________________ organisms in the ecosystem produce.
Answer: photosynthetic Difficulty: II Section: 2 Objective: 1

93. In an ecosystem, ____________________ diminishes at each successive trophic level.
Answer: energy Difficulty: I Section: 2 Objective: 1

94. A path of energy through the trophic levels of an ecosystem is called a(n) ____________________ ____________________.
Answer: food chain Difficulty: I Section: 2 Objective: 2

95. The interrelated food chains in an ecosystem are called a(n) ____________________ ____________________.
Answer: food web Difficulty: I Section: 2 Objective: 2

96. Decomposers are part of a special class of consumers called ____________________.
Answer: detritivores Difficulty: II Section: 2 Objective: 2

97. An energy pyramid shows the amount of energy contained in the bodies of organisms at each ________________ level.
Answer: trophic Difficulty: I Section: 2 Objective: 2

98. Every time that energy is transferred in an ecosystem, potential energy is lost as ________________.
Answer: heat Difficulty: I Section: 2 Objective: 3

99. When forests are cut down, both water and nutrient ________________ are altered.
Answer: cycles Difficulty: I Section: 3 Objective: 1

100. Water that seeps into the soil is called ________________ water.
Answer: ground Difficulty: I Section: 3 Objective: 1

101. Carbon is returned to the atmosphere by cellular respiration, combustion, and ________________.
Answer: erosion Difficulty: I Section: 3 Objective: 2

102. The conversion of nitrogen gas to ammonia by the action of bacteria is called ________________ ________________.
Answer: nitrogen fixation Difficulty: II Section: 3 Objective: 3

103. Nitrogen ________________ is the absorption and incorporation of nitrogen into plants and animals.
Answer: assimilation Difficulty: II Section: 3 Objective: 3

104. The process of ________________ occurs when anaerobic bacteria break down nitrates and release nitrogen gas back into the atmosphere.
Answer: denitrification Difficulty: II Section: 3 Objective: 3

PROBLEM

Nitrogen fertilizer is added to soils in virtually all agricultural areas of the world. The use of nitrogen fertilizer greatly increases the amount of food produced. However, it can also affect the ecology of areas near agricultural areas. The data presented in the table below were obtained in an experiment conducted to evaluate the effects of nitrogen fertilizer on grass species diversity. Nitrogen fertilizer was applied yearly to an experimental plot, beginning in 1856.

Year	1856	1872	1949
Total number of grass species	49	15	3
Number of species producing more than 10% of the total dry weight of all species combined	2	3	1
Number of species producing more than 50% of the total dry weight of all species combined	0	1	1
Number of species producing more than 99% of the total dry weight of all species combined	0	0	1

105. Refer to the experiment and data table above. Write three conclusions that you can draw from these data.
Answer: The following are some possible conclusions:
(1) The total number of grass species decreased over time and with exposure to nitrogen fertilizer. (2) At the beginning of the experiment, there was no one dominant species of grass. Over time and with exposure to nitrogen fertilizer, a few species became dominant. (3) Prolonged use of nitrogen fertilizer encourages the growth of one or a few dominant species.
Difficulty: III Section: 1 Objective: 2

106. Refer to the experiment and data table above. How could this experiment have been designed differently to make it a better experiment?

Answer:

It should have been designed to have a control plot that did not receive nitrogen fertilizer. As the experiment was designed, the effects of nitrogen fertilizer cannot be distinguished from the effects of time.

Difficulty: III   Section: 1   Objective: 2

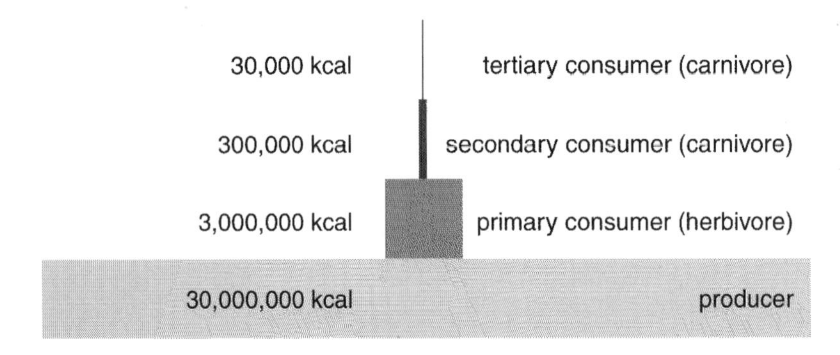

The diagram above depicts a typical energy pyramid. It shows how energy is lost as it is transferred from one level to another in the food chain of the pyramid.

107. Refer to the illustration above. Explain what becomes of energy in organisms as it is transferred up through a food chain.

Answer:

As energy is passed from one level of a food chain to another, only about 10 percent of it is available to the next organism up in the food chain. This occurs in part because all energy conversions involve a loss of energy in the form of heat. It is also partly due to the fact that each organism in a food chain requires energy to maintain its body's activities. Thus, the amount of useful energy to do work and support life processes decreases as energy passes through trophic levels.

Difficulty: III   Section: 2   Objective: 3

ESSAY

108. Explain how a change in the habitat of a species affects the entire ecosystem.

Answer:

A change in habitat may disturb the interactions of plants and animals in the ecosystem. A drastic change in the factors of a habitat that affects one species can have an effect on the whole ecosystem because it affects the natural cycling of nutrients, food chains, and food webs. This disruption could result in endangerment or extinction of species in the ecosystem.

Difficulty: II   Section: 1   Objective: 1

109. What is the difference between primary and secondary succession?

Answer:

Primary succession is the replacement of species in an area that had not previously supported life, such as bare rock or a sand dune. Secondary succession involves species replacement in habitats that have been disrupted due to natural disaster or human activity but that still possess a small amount of soil and vegetation.

Difficulty: II   Section: 1   Objective: 3

110. Explain how alders can lead to the destruction of a sturdy plant like *Dryas*.

Answer:

Alders grow faster than *Dryas*, and dead leaves and fallen branches from them add useable nitrogen to the soil. The added nitrogen allows other trees to invade and develop vigorously. Eventually, dense thickets of the trees shade and kill the *Dryas*.

Difficulty: II Section: 1 Objective: 3

111. Clover plants, rabbits, and coyotes are some of the organisms that occupy a particular ecosystem. Assign the roles of primary producers, primary consumers, and secondary consumers to these three groups of organisms and explain your answer.

Answer:

In this ecosystem, the clover plants are the primary producers. They help manufacture the organic nutrients necessary to sustain the ecosystem. Rabbits are herbivores that consume the primary producers (the clover plants), so they are classified as primary consumers. Coyotes eat the primary consumers (the rabbits), so they are classified as secondary consumers.

Difficulty: II Section: 2 Objective: 1

112. Why are decomposers necessary for the continuation of life on Earth?

Answer:

Decomposers release matter from waste materials and dead organisms. Were it not for the action of decomposers, the Earth would eventually be depleted of usable essential matter such as carbon and nitrogen, that organisms need. Without decomposers, essential materials would not be recycled.

Difficulty: II Section: 2 Objective: 1

113. Describe how energy is transferred from one trophic level to another.

Answer:

A portion of the energy available to the organisms at each level of the food chain is stored in the chemical bonds of nutrients that are not used by an organism in order to sustain life. When that organism is eaten by another, the stored chemical energy is transferred to the new organism and used to sustain its life.

Difficulty: II Section: 2 Objective: 3

114. Why is it cheaper for a farmer to produce a pound of grain than a pound of meat?

Answer:

Animals are on higher trophic levels than plants. Consequently, it takes more energy to produce one pound of meat than to produce several pounds of grain.

Difficulty: II Section: 2 Objective: 3

115. A plant disease infects most of the vegetation in a particular area, destroying it. How might the destruction of this vegetation affect the animal life in the area?

Answer:

The ecosystem would be seriously disrupted. Herbivores that ate the vegetation would be affected if this was their major source of food. The carnivores in the area would soon die or leave the area because their source of energy—the herbivores—could not remain.

Difficulty: II Section: 2 Objective: 3

9997265947 1 2 3 4 5 6